The Future of Prevention and Treatment of Breast Cancer

Jose Russo

The Future of Prevention and Treatment of Breast Cancer

Springer

Jose Russo
Breast Cancer Res Lab, Room P 2037
Fox Chase Cancer Center
Philadelphia, PA, USA

ISBN 978-3-030-72814-4 ISBN 978-3-030-72815-1 (eBook)
https://doi.org/10.1007/978-3-030-72815-1

This Springer imprint is published by the registered company Springer Nature Switzerland AG
The registered company address is: Gewerbestrasse 11, 6330 Cham, Switzerland

Preface

The objective of this book is to provide an analysis of the present and future prevention and treatment strategies of breast cancer, emphasizing the cost benefits and quality of life of the patient when possible.

Rooted in the present knowledge in breast cancer biology and prevention and treatment options, the book extrapolates the future tools that could be available for oncologists and how these new approaches could change the landscape of recurrence or survival of the disease.

Special emphasis will be given to prevention strategies counterposing the present limitations and conflict prevention guidelines for both hereditary and non-hereditary breast cancer. The implementation of new preventive strategies based on the present knowledge could save lives and be more cost efficient.

The book presents a critical review of the treatment and prevention of breast cancer and how a quantum leap could be achieved by applying current basic research knowledge in the clinic.

Philadelphia, PA, USA

Jose Russo

Contents

Contributors

Yanrong Su The Irma H Russo, MD-Breast Cancer Research Laboratory at the Fox Chase Cancer Center-Temple Health, Philadelphia, PA, USA

Magda J. Vandeloo The Irma H Russo, MD-Breast Cancer Research Laboratory at the Fox Chase Cancer Center-Temple Health, Philadelphia, PA, USA

Chapter 1
Defining Breast Cancer

Abstract The name of "breast cancer" is the general term for carcinomas of the breast tissue, and it can be presented in young as well as in older women. Several criteria are used to classify breast tumors; these include histologic pattern, degree of nuclear pleomorphism, number of mitoses, presence of inflammatory response, blood vessel invasion, lymph node involvement, and the presence of hormone receptors. Using these basic morphologic criteria, tumors of the breast have been classified in benign and malignant and among the latter are sub-classified in in situ and invasive lesions. In this chapter and in the entire book we are referring to breast cancer as the invasive variant of tumor. In the classical diagnostic tools of the pathologist, the invasive tumors were additionally classified according to the pattern visualized under the microscope such as invasive cribriform, mucinous, tubular, medullary, lobular, etc. The rationale for a genomic classification is, basically, two: one is that knowing the gene expression profile of a tumor leads us to understand cancer behavior and second at the clinical level can help us to identify genes that are associated with specific cancer phenotype. The modern, accepted subtypes are the luminal A, luminal B, basal-like, ERBB2+/HER2+, and the normal breast-like subtypes. There are several accepted means for distinguishing between the five subtypes. The primary method is the presence or absence of three different cellular receptors in the breast cancer tumors. The three receptors are the estrogen receptor (ER), the progesterone receptor (PR), and the human epidermal growth factor receptor 2 (HER2). The protein Ki-67, a known prognostic factor associated with proliferation, is utilized in a similar manner, looking at the low or high levels for subtype distinction. The grade of the tumor is another factor incorporated into identifying molecular subtypes, which looks at how the tumors appear in comparison to normal, well-differentiated breast tissue. High grades are described as "poor," or not well-differentiated, while low grades are described as "good," or well-differentiated. Additional classifications have set out to further distinguish between these accepted subtypes. Several groups have sought out newer, more reliable studies for means of classifying breast cancers. Among them are the studies done by the Molecular Taxonomy of Breast Cancer International Consortium (METABRIC) and the Normal Cell Subtype-based classification system, where they focused on utilizing normal cell types found in normal breast tissue as references for breast cancer classifications. After the publication of the new genomic classification of breast cancer, several adaptations compiling few representative genes have been introduced in the practice of oncology. The MammaPrint contains 70 gene expression signature, and

J. Russo, *The Future of Prevention and Treatment of Breast Cancer*,
https://doi.org/10.1007/978-3-030-72815-1_1

the major trial PAM or prediction analysis of microarrays recapitulated the microarray classifier using RT-PCR-based PAM50 assay. PAM50 correlates well with the Oncotype and the use of the four IIC parameters (ER, PR, HER2, and Ki-67). The emergence of better and faster DNA sequence analysis has allowed to characterize the mutations of known as well as less explored parts of the genome such as base substitutions, small insertions/deletions, rearrangements, and copy number changes.

Keywords In situ · Invasive breast cancer · Luminal A · Luminal B · Basal-like · ERBB2+/HER2+ · Normal breast-like subtypes. METABRIC · Normal Cell Subtype-based classification system · MammaPrint. PAM50 · Oncotype

1.1 Introduction

The name of "breast cancer" is the general term for carcinomas of the breast tissue. Breast carcinomas are heterogeneous and vary greatly in physical appearance, or their phenotype, both at macroscopic and molecular scale. This phenotypic diversity corresponds to variation in gene expression producing a classification for breast cancer that more accurately predicts tumor behavior. The phenotypic diversity will be described in Sect. 1.2, and a classification of tumors based on their genotype will be described in Sect. 1.3.

Breast cancer can be presented in young as well as in older women. The criteria for defining a young woman were published in a study [1] using the Surveillance, Epidemiology, and End Results database of all cases of female breast cancer diagnosed between 2000 and 2007. This study has used age-specific gene expression profiles such as RNA sequence data from the Cancer Genome Atlas database. The age of 40 years was determined as the optimal cutoff value. Among 94,087 patients, 12,755 were aged 40 years or younger (younger group), and 81,332 were older (older group). The 5- and 10-year cancer-specific survival rates in younger and older groups were 88.74% and 80.65%, respectively, and 93.22% and 88.43%, respectively ($P < 0.001$). Young patients with breast cancer, especially those in the ER+ subgroup, have worse prognosis in this study [1].

1.2 Invasive Breast Cancer

Several criteria are used to classify breast tumors; these include histologic pattern, degree of nuclear pleomorphism, number of mitoses, presence of inflammatory response, blood vessel invasion, lymph node involvement, and the presence of hormone receptors [2–16]. Using these basic morphologic criteria, tumors of the breast have been classified in benign and malignant and among the latter are sub-classified in in situ and invasive lesions. In this chapter and in the entire book, we are referring to breast cancer as the invasive variant of tumor.

Invasive breast cancer includes all the tumors in which stromal invasion is detectable including the microinvasive carcinoma. The vast majority of breast carcinomas are described as invasive ductal carcinomas based on architectural patterns and cytological features. This invasive ductal carcinoma is also called not otherwise specified (NOS) comprising 70–80% of the primary breast cancer. The remainder of the invasive carcinomas are classified as lobular, tubular, medullary, etc. as depicted in Table 1.1. A detailed description of all these types has been done in a previous publication [16]. In this chapter I will summarize the basic characteristics of each of the invasive breast cancers.

The most frequent breast cancer is the so-called NOS. These tumors are variable in size and shape, firm, and poorly circumscribed. They may also contain areas of necrosis, hemorrhage, and cystic degeneration. Histologically the cells grow in diffuse sheets or in well-defined nests or cords or as individual cells. The cancer cells are mostly positive for different types of keratin type 8, 18, and 19. Seventy percent of cases are positive for lactalbumin and other markers like CEA [17], B72.3, and BCA-225 [18–21]. Epithelial membrane antigen is a good marker of breast epithelial cell, as well as E-cadherin and P53.

We have reported [22] as well as others [23, 24] that breast carcinomas can be immune-reactive for S-100 protein. The basement membrane components laminin and collagen IV show a fragmented pattern of expression. Vimentin can be expressed in certain areas of the tumor and may be an indication of EMT or Epithelial Mesenchymal Transition [25]. Marker of stem cells like CD44 is also shown in breast cancer. Marker for desmin is frequently found in the stromal component of

Table 1.1 Types of invasive carcinoma of the breast

Invasive ductal with an extensive
Intraductal component
Invasive lobular
Mucinous
Medullary
Papillary
Tubular
Adenoid cystic
Secretory
Apocrine
Cribriform
Paget's disease of the nipple
With invasive carcinoma
Without invasive carcinoma
Carcinoma with metaplasia
Squamous type
Spindle cell type
Cartilaginous and osseous type
Mixed type
Inflammatory

the tumor, whereas markers for smooth muscle antigen (SMA) can be present in remnants of myoepithelial cells as well as in the wall of blood vessels [16]. Whereas the light microscopy and immunocytochemistry solve most of the cases of breast cancer, there are undifferentiated tumors in which the electron microscope is helpful in identifying luminal spaces border by microvilli that indicate the glandular origin, and for more details on the use of the electron microscope in tumor diagnosis, see Reference [26].

There are reports in the literature [27] that indicated that if the cancer cells are arranged in large nests, their metastatic potential is decreased when compared with an architectural pattern formed by numerous small nests of cancer cells.

In the classical diagnostic tools of the pathologist, the invasive tumors were additionally classified according to the pattern visualized under the microscope and were called invasive cribriform carcinoma when they have a similar appearance to that seen in ductal carcinoma in situ but exhibiting stromal invasion, and this type has been associated with an excellent prognosis [28, 29]. In some other occasions there are small clusters of tumor cells surrounded by extracellular mucin [30, 31]. Pure mucinous carcinoma is associated with a very low incidence (2–4%) of nodal metastases [32–34].

Another pattern of invasive breast cancer is the one called tubular that is in general small, with a mean diameter of about 1 cm [35, 36]. This type of tumor is characterized by the prominent tubular arrangement of the cancer cells surrounded by prominent stroma of collagen, that is, some areas are hyaline and positive for amyloid. The glandular spaces are irregular and often angulated, and they lack myoepithelial cell component and basement membrane. Metastases to axillary nodes occur at very low frequency (10% of cases), and the prognosis is excellent [35, 37–39].

The so-called medullary carcinoma is well circumscribed, and the borders are always of the "pushing" type. The pattern of growth is diffuse, with minimal or no glandular pattern. The tumor cells are large and pleomorphic, with large nuclei and prominent nucleoli and numerous mitoses. The cell borders are indistinct, giving the tumor a syncytial or sheet-like appearance. An important feature of this type of tumor is the prominent infiltrate of lymphocytes and plasma cells. The prognosis for medullary carcinoma is better than for the ordinary invasive ductal carcinoma [40, 41].

Other types include the invasive papillary, the apocrine [42], the juvenile (secretory) carcinoma [43–45], and those cancers with neuroendocrine features. Much less frequent are also the metaplastic carcinomas [17] that are in general more aggressive than the ordinary invasive ductal-type carcinoma [46, 47].

The inflammatory carcinoma is a clinical term for a type of breast carcinoma in which the entire breast is reddened and warm, with widespread edema of the skin, thus simulating the appearance of mastitis. Most of this type of breast cancer are undifferentiated carcinomas with widespread carcinomatosis of the dermal lymphatic vessels.

Among the invasive carcinomas of the breast originated in the lobules type 2 of the breast are invasive lobular carcinomas (ILC) and characterized by the presence of small and relatively uniform tumor cells growing singly, in Indian file, and in a concentric or "targetoid" fashion around lobules that usually are involved by in situ lobular neoplasia. The stroma of this type of tumor is formed by dense fibrous type containing foci of per ductal and per venous elastosis. These tumors react positively to heavy molecular weight keratin, are p53 negative, and most importantly have a decrease or absence of E-cadherin [48–51]. A variety of lobular invasive carcinoma is the pleomorphic lobular carcinoma that has the pattern of growth of a classical breast carcinoma but exhibits a marked degree of nuclear pleomorphism and abundant cytoplasm [52, 53]. It also frequently shows apocrine differentiation, focal signet ring morphology, lack of hormone receptors, higher expression of p53 and *HER2/neu,* occasional expression of chromogranin, and lack of E-cadherin staining [54–56]. Another variant of lobular invasive carcinoma is the tubule-lobular carcinoma that is characterized by the admixture of small tubular formations having a minute or undetectable lumen ("closed" or "almost closed" tubules) with cords of tumor cells growing in a lobular configuration similar to that of invasive lobular carcinoma [57].

Two other breast cancer types described in the literature are the microinvasive carcinoma and Paget's disease. Microinvasive carcinoma is defined as any carcinoma in situ of the breast showing one or more areas of stromal invasion not surpassing 1 mm in thickness. Immunohistochemical evaluation with myoepithelial and basement membrane markers is useful for a confirmation of the diagnosis. Overall patients with microinvasive carcinoma are at risk for nodal metastases, but their survival rate is better than for patients with T1 invasive carcinoma [58]. Paget's disease is the name given to eczema-like lesions centered in the nipple caused by breast carcinoma [59]. It is most of the time accompanied by an underlying breast carcinoma of in situ ductal type, with or without associated stromal invasion. Under the microscope the tumor is characterized by large clear cells with atypical nuclei seen within the epidermis, usually concentrated along the basal layer but also permeating the Malpighian layer.

1.3 Breast Cancer Genotypes

The rationale for a genomic classification are basically two: one is that knowing the gene expression profile of a tumor leads us to understand cancer behavior and second at the clinical level can help us to identify genes that are associated with specific cancer phenotype. Knowing the genomic signature of a given breast cancer provides the molecular pathways that can help better targeting of available therapy.

The first distinguishing classification came from the study performed by Perou and Sorlie [60]. In this study they acquired samples of breast tissue from 42 individuals. Forty of these were breast cancers (mostly invasive ductal carcinomas), one was a fibroadenoma, and the last sample normal breast tissue. In addition, contained in this study are 22 pairs of tumor samples from which 20 were paired before and after a chemotherapy regimen, and 2 pairs from primary tumors paired with their respective lymph node metastasis. Using these samples, they isolated the RNA and performed cDNA microarray. From these microarrays 8102 genes were initially identified, and a subset of these genes was selected based on the variation in their expression using at least plus-or-minus four-times the median level of expression. Using these criteria, they finally selected 1753 for hierarchical cluster for their final classification. They selected genes based on gene expressions that were similar in any sample taken from the sample tumor, but varied between different tumors, and for that purpose they utilized the 22 paired tumor samples and identified 496 genes from the 1753 identified earlier based on the variation in their gene expression, with the added distinction of having greater variation between different tumors than from the same tumor. They called this new cluster the "Intrinsic Gene Subset." From this new subset, they were not only able to determine the expression levels for each sample, but were also able to group them based on their expression within the two layers found in mammary gland structures (the lobules and ducts): the inner luminal epithelial cells and the surrounding basal myoepithelial cells. From these groupings, they were able to further distinguish each group into gene clusters. They identified one cluster for the luminal epithelial cells, the luminal epithelial/estrogen cluster. They also identified three clusters for the basal epithelial cells: the ERBB2 overexpression cluster, the basal epithelial cell-associated cluster, containing keratins 5 and 17, and the basal epithelial-cell-enriched gene cluster [1]. The clusters identified were later refined into subtypes [61]. The luminal cluster was separated into luminal subtype A, luminal subtype B, and luminal subtype C. The basal clusters were redefined: the ERBB2 overexpression cluster became simply the ERBB2+ subtype, the basal epithelial cell-associated cluster became the basal-like subtype, and the basal epithelial-cell-enriched gene cluster became the normal breast-like subtype. From these produced subtypes, they also looked at the clinical features indicated by each subtype. To accomplish this, they utilized data acquired from 49 breast cancer patients showing all 5 subtypes, whom only had diseases local to the breast and with little-to-no metastasis present [61]. They specifically looked at the overall survival (survival months) and relapse-free survival (RFS) probabilities for each subtype over a 4-year period, in comparison to the other subtypes. In addition, they also looked at the outcomes when luminal subtype C and B were grouped with the other subtypes. Analysis showed that the basal-like and ERBB2+ subtypes both had the lowest RFS and overall survival. Additionally, the luminal subtype C was shown to have the worst overall survival of all the luminal subtypes, and subtypes ERBB2+ and luminal B were shown to share certain genes associated with a poor prognosis.

1.3.1 Molecular Subtypes of Breast Cancer

Since the defining of the intrinsic subtypes, many studies have gone on to further refine and expand upon the initial breast cancer classifications of Perou and Sorlie [60, 61], redefining them as the Molecular Subtypes of Breast Cancer. The modern, accepted subtypes are the luminal A, luminal B, basal-like, ERBB2+/HER2+, and the normal breast-like subtypes. There are several accepted means for distinguishing between the five subtypes. The primary method is the presence or absence of three different cellular receptors in the breast cancer tumors. The three receptors are the estrogen receptor (ER), the progesterone receptor (PR), and the human epidermal growth factor receptor 2 (HER2). Overexpression of these receptors has been observed in breast cancers, but have often only been looked at individually. In addition, different breast cancer tumors have been shown to have different expression levels of these receptors. Thus, by looking at all the receptors together and identifying which are overexpressed (or absent) in the tumor cells, there can be a clear classification used to distinguish between breast cancers.

The protein Ki-67, a known prognostic factor associated with proliferation, is utilized in a similar manner, looking at the low or high levels for subtype distinction. The grade of the tumor is another factor incorporated into identifying molecular subtypes, which looks at how the tumors appear in comparison to normal, well-differentiated breast tissue. High grades are described as "poor," or not well-differentiated, while low grades are described as "good," or well-differentiated. A summary of these features in the molecular subtypes can be seen in Table 1.2 [62, 63].

Additional classifications have set out to further distinguish between these accepted subtypes. This includes the further subtypes for luminal B: the HER2+ and HER2- subtypes. The distinguishing factor between them is that Ki-67 levels are generally high in HER2+ luminal B breast cancers [62]. Another distinction is made between triple-negative breast cancers (TNBC) and basal-like breast cancers. Though TNBC (named for being negative for all three receptors) have traditionally been grouped with the basal-like subtype, they are not synonymous, and there is at most an 80% overlap between the two [63]. TNBCs have thus been separated into two further subtypes: basal-like and non-basal-like. The major distinction is the expression of cytokeratin 5 and 6 (CK5/6), as well as epidermal growth factor receptor (EGFR), for the basal-like subtype [64].

Other studies have set out to describe additional molecular subtypes, distinct from the accepted five. One proposed subtype is the claudin-low subtype. They are characterized with low expression of the claudin proteins (found within cellular junctions) and are associated with mammary stem cells [65]. Although similar to the basal-like subtype in being triple-negative, the claudin-low subtype clinically shows a better prognosis than the basal-like subtype.

Table 1.2 Summary of the standard features for each of the five molecular subtypes [62, 63]

Molecular subtype	Estrogen (ER)	Progesterone (PR)	HER2	Ki-67	Tumor grade
Luminal A	+and/or	+	–	Low	Low
Luminal B	+and/or	+	+/-	High (HER2+)	High
ERBB2+/ HER2+	–	–	+	High	High
Basal-like ("triple-negative")	–	–	–	High	High
Normal breast-like	Normal	Normal	Normal	Normal	Low

Based on the immunocytochemical classification, the luminal A-subtype breast cancers suggest that these patients have a better prognosis compared with those with breast cancers of other subtypes. In a publication of Gao and Swain [66], they raised the question that in these patients chemotherapy could be omitted and endocrine therapy alone could be sufficient based on the fact that the luminal A-subtype tumors are a unique subset that may have favorable tumor biology [66].

1.3.2 The Molecular Taxonomy of Breast Cancer International Consortium (METABRIC) Classification

Despite the usefulness of the molecular subtype classifications of breast cancer, there are several limitations. One of the major limitations is the apparent lack of understanding of the variation in response to therapies specific to the subtype. Such variation has limited value in a clinical setting, where proper treatment is crucial to patient survival. Because of these limitations, several groups have sought out newer, more reliable studies for means of classifying breast cancers. Among them are the studies done by the Molecular Taxonomy of Breast Cancer International Consortium (METABRIC) [67]. The methodology used by this consortium used sequencing technologies that identified the mutational patterns and genomic instabilities characteristic of different breast cancers. This new classification also set out to integrate the classical classifications of breast cancer, describing features such as receptors and tumor grade, as well as direct comparisons with the molecular classifications. In the study, about 2000 breast tumors were analyzed, to acquire both their genomic and transcriptomic sequences, identifying where gene alterations had occurred. This included inherited variation to the genome, specifically single nucleotide polymorphisms (SNPs) and copy number variants (CNVs), but also looked at acquired variation via single nucleotide variants (SNVs aka mutations) and copy number aberrations (CNAs) [67]. From this data, they were able to produce ten novel subtypes of breast cancer, which they designated Integrative Clusters (IntClust 1–10). Each cluster was primarily distinguished by the CNAs, which were identified to have the greatest variation, but also found to have overall differing gene expressions. The extent to which the clusters associated with the accepted intrinsic subtypes was analyzed for each cluster, as was the expression of the accepted prognostic receptors of estrogen, progesterone, and HER2. Further analysis identified the clinical implications for each cluster, such as the genomic instabilities, and distinguishing somatic mutations, but also more specific characteristics for each cluster including age of diagnosis and survivability probabilities.

1.3.3 *Genomic Classification Based on the Normal Cell Subtype*

Despite the benefits of the newer genomic classifications of breast cancer, alternative means of classifications still arise to confront new or unaddressed issues. Such issues were addressed in a study by Santagata et al. [68]. In their study, they set out to produce a Normal Cell Subtype-based classification system, where they focused on utilizing normal cell types found in normal breast tissue as references for breast cancer classifications. This method, they argue, has successfully been used before to characterize hematopoietic tumors (lymphomas, leukemias, etc.) by other research groups [69], but have rarely been emulated due to a poor understanding of cell-type diversity among tissues. They argue that their new classification, unlike previously produced ones, forms actual disease taxonomy for breast cancers. That is, previous classification systems have heavily relied on differing clinical results (based on different molecular platforms for analysis) to form categories based only on overall prognosis. These categories also vary greatly; no new classification system is truly agreed upon in clinical settings, seeing them as unreliable for patient prognosis and treatments. Their new classification system aimed to provide such clinical reliability. This classification identified that breast cancers, being heterogeneous, can vary depending on their cellular origin: in the luminal epithelial layers or the myoepithelial layers [67]. Thus, they analyzed about 15,000 normal breast cells for cellular markers distinguishing between the two layers. They focused on identifying bimodal expression markers (which produced a clear negative/positive distinction), and utilizing these markers to distinguish between varying differentiation states of the cell populations [68]. Three of the major markers identified were hormone receptors of the luminal epithelia: the vitamin D receptor (VDR), the androgen receptor (AR), and the estrogen receptor (ER). Additional markers included different keratins, claudins, cluster of differentiation (CD) markers, and even Ki-67. Identifying the different expression of these markers in the different cell populations allowed for the formation of eleven luminal layer subtypes (L1–11) and 2 myoepithelial layer subtypes (My1 and My2). Following these classified layers, the study focused on actually classifying human breast tumors based on normal cell types. Four unique subtypes, called "Hormone Receptor Subtypes," were identified: HR0, HR1, HR2, and HR3. Each subtype is based on the expression of the three major hormone receptors (VRD, AR, and ER) and how many were expressed (0–3). The previously characterized luminal subtypes were then distinguished based on these novel subtypes. Next, the study looked to identify if breast tumors maintain the same expression patterns characteristic of the normal cell type, specifically the differentiation state-specific patterns; this involved identifying the gene expression patterns among the luminal and basal markers (including the three major markers), as well as the specific marker of K5/K14 (found to be a reliable distinguisher between luminal layers). The expression of these markers was identified in ER+, HER2+, and triple-negative breast cancer (TNBC) tumors and compared with the expressions found in normal breast tissues with the same distinguished expressions. An example of this comparison can best be seen for the ER+ tumors, where they identified the ER+ tumors to co-express VDR in 93% of the tumors and AR in 59% of the tumors, and the K5/K14 were found to be negative in these tumors. When compared to the counterpart normal cells, there was found to be a near identical expression pattern: they both co-express VDR and AR to the same levels, and both rarely expressed K5 or K14. Such identical expressions were seen in the HER2+ and TNBC tumors and their counterparts, verifying their Normal Cell Subtype Classification method [68]. Finally, the study aimed to identify the clinical significance behind the new Hormone Receptor subtypes. This involved acquiring tumor data from patients in a separately performed study by the Nurse' Health Study, which had previously been classified as ER+, HER2+, and TNBC based on the presence of the classical receptors (ER, PR, and HER2) within their tumors. These classically assigned tumors were compared with the new HR subtypes, which showed that the HR subtypes provided more distinguished groups of tumors than the previous classification. In addition, the HR subtypes were clinically identified from each other by overall survival and relapse-free survival, which identified that the HR0 subtype had the worst prognosis and the HR3 subtype had the best prognosis [68].

1.4 Prognostic Markers

Using the tumor histologic pattern, degree of nuclear pleomorphism, number of mitoses, presence of inflammatory response, blood vessel invasion, lymph node involvement, and the presence of hormone receptors [3–5, 7, 9, 10, 12–14, 69–85], attempts have been made to identify patients at high risk of early disease recurrence and thus to more effectively target aggressive adjuvant chemotherapy and intensive follow-up protocols. However, many breast carcinomas having similar histologic, nuclear, or mitotic grades or estrogen receptor (ER) levels differ markedly in their recurrence behavior, thus affecting a patient's survival. It is therefore obvious that single morphologic or biologic characteristics are insufficient to predict the biologic behavior of a tumor; therefore, a combination of various criteria may be necessary to accurately separate subpopulations of patients with breast cancer at increased risk of recurrence or shortened survival [70]

One of the advantages of the genomic classification is that it has provided in addition to the diagnosis a therapeutic approach that is of great significance for the practical oncologist. The genomic classification also has provided not only a therapeutic conduct but also has provided a reliable way to determine the prognosis of the disease. More studies in this direction are coming when better delineation of the genes and the therapy is established. Therefore, it is still important to bear in mind which are the prognostic markers that are still reliable and that not only can be used in combination with the genomic markers such as the *METABRIC* but also could still be used as another independent parameter for evaluating therapeutic success.

A major focus of this section is how to use single characteristics of primary breast tumors, such as ER status, histologic grade (HG), nuclear grade (NG), mitotic grade (MG), and the combinations of HG, MG, and NG to give a final grade (FG) similar to that of Bloom and Richardson [4] and lymph node status (LN) of the patient at the time of surgery to identify subpopulations of patients at increased risk of early disease recurrence or shorter survival.

1.4.1 *Tumor Grading*

Tumors in general are graded according to the criteria established by Bloom and Richardson [4, 86] and selecting for grading the least differentiated area of the tumors, instead of the average, in order to determine the weight that the most undifferentiated areas have in tumor prognosis (Table 1.3). The nuclear grade 1 is characterized by nuclei that are uniform in size and shape and the chromatin is uniformly granular. The nuclear grade 2 is characterized by a moderate variation in size and shape, with mild chromatin condensation and rarely the nucleolus is prominent. The nuclear grade 3 is characterized by marked degree of pleomorphism by variation in size and shape. The nuclear grade 3 is also characterized by having a more irregular chromatin and a prominent nucleolus. The histological grade is determined by the conservation of the ductular or glandular pattern. In the histologic grade 1 most of the tumor is formed by well-preserved ductular pattern, whereas in histological grade 2 there is a moderate formation of tubules and solid masses of tumor cells that can be intermingled with other containing lumen formation. When the tubular or glandular formation has been completely lost, the histological grade 3 is given. The mitotic grade is determined as indicated in Table 1.3 by the number of mitosis per 10 high power fields (HPF) with the adequate correction for the specific microscope, since there are small variations depending on the brand. It is accepted to count mitosis with a 40× objective and a 10× ocular, with a field area of 0.152 mm^2. The most important is to recognize the mitotic figures in the different phases of the cycle to have a consistent counting of mitosis. Abnormal mitoses should also be counted.

The Final Grade or FG is determined by adding the values obtained for HG, MG, and NG. A total score of 3–5 represented an FG of I; 6–7, an FG of II; and score of 8–9, an FG of III (Table 1.3) [86]. The higher mitotic grade is usually seemed in tumors with nuclear grade 3 however this is not a rule.

Table 1.3 Criteria for tumor grading

Tubule formation	Nuclear pleomorphism	Mitotic count for a 40× objective with a field area of 0.152 mm^2
Majority of tumors >75% (score 1)	Small regular nuclei (score 1)	0–5 mitoses per 10 HPF (score 1)
Moderate 10–75% (score 2)	Moderate increase in size (score 2)	6–10 mitoses per 10 HPF (score 2)
Minimal <10% (score 3)	Marked variation in size, nucleoli, chromatin clumping (score 3)	>10 mitoses per 10 HPF (score 3)

1.4.2 Lymphocytic Infiltration

It is well known that tumor-infiltrating lymphocytes are associated with the improved survival in different cancer types [87–90] and in breast cancer have been shown to be a good predictor of complete remission after preoperative chemotherapy [91–95]. Patients with breast cancer with the triple-negative subtype associated with abundant TILs showed improved rates of survival [96–98]. Breast cancers in which the density of lymphocytes or plasma cells is greater than 50–60% are called lymphocyte-predominant breast cancer. However, the main problem is the lack of agreement among pathologists to evaluate lymphocytic infiltration even though a working classification has been published [99].

It has been reported in the literature [100] that $CD8^{+}$ T cells [101] or cells positive for cytotoxic granules (granzymes or perforin) contributed to better response to chemotherapy [93], an indication of cytotoxic immune responses. The St Gallen Consensus meeting in 2015 reported that tumor infiltrate lymphocytes do not have neither prognostic nor predictive marker in breast cancer [102] although in a recent publication [103] it has provided evidence of its usefulness.

1.4.3 ER and PR as Biomarkers of Prognosis

The standard evaluation of breast cancer for clinical purposes involves IHC characterization of ER, PR, and HER2 status. Hormone receptor-positive breast cancers account for around 75–80% of all cases, and standardized IHC assays for the routine testing of ER and PR are used to guide the selection of patients for hormonal-based therapies. HER2 represents the only additional predictive marker currently in routine use, and 10–15% of breast cancers have HER2 overexpression and/or amplification with around half of these co-expressing hormone receptors. The remaining 10–15% of breast cancers are negative for the three hormone receptors and have been called triple-negative breast cancer.

Estrogen receptor has, for a long time, been known to play a crucial role in the carcinogenesis of most breast cancers. More than 50% of breast cancers have an abundance of estrogen receptors, particularly ER-alpha, the presence of which correlates with a positive prognosis for the patient. The absence of an overexpression of ER-alpha has the inverse prognosis [101]. The effects of estrogens are primarily carried out by their interaction with estrogen receptors, which are part of a large super family of nuclear membrane bound, ligand-activated transcription factors [104]. Estrogen receptor expression is present in 10–20% of normal human mammary tissue and plays an important role in physiological events like puberty and pregnancy. Estrogen receptors come in two isoforms, ER-alpha and ER-beta, both of which appear in homology in breast cancer, yet play a different role [105]. It has also been found that the tissue distribution and function of both ER-alpha and ER-beta are similar in rats as they are found in human. Both receptors have similar binding regions, but have different gene expression characteristics [106]. ER-beta is mostly involved in the inhibition of tumor growth [105]. Estrogen receptors have two paths of function, both of which regulate gene expression, direct and indirect [107]. The direct, or classical pathway, is when a ligand (estrogen or anti-estrogens) activated ER binds cis-regulatory elements on the DNA called estrogen response elements (EREs) [106]. The indirect pathway is where the ER binds transcriptional factors to the N-terminal activating region, and then the transcriptional factors bind the DNA; this is called transcriptional crosstalk [108]. In both cases the estrogen receptors dimerize and interact with a co-regulator complex that help make the DNA accessible from the chromatin [106]. Estrogen receptors may also have no genomic actions by interacting with scaffold proteins or various signaling molecules [108]. In breast cancer cells some no genomic actions of an estradiol activated ER-alpha include activation of the MAP-K pathway, interaction with ErbB2 (HER-s/neu), and activation of epidermal growth factor receptors, all of which are involved in cell proliferation [108].

The molecular composition of mammary carcinomas plays a deciding role in the form of treatment to be used. Today a more precise way of classifying human breast cancer is to use molecular diagnostics [107]. Simply diagnosing based on morphological features can be inaccurate and lead to misdiagnoses because many tumors might deviate from the typical morphology of that type of carcinoma [107]. Breast cancer is typically divided into two main groups: estrogen receptor positive, which includes luminal A and B subtypes, and estrogen receptor negative, which includes HER2, normal-like, and basal-like cancers [109]. Depending on the expression of these markers, as well as other genes and proteins, the patient will receive a different therapy regimen. Once the estrogen/progesterone status is determined, a course of treatment can be mapped out. If the carcinoma is positive for these receptors, then endocrine therapy is an option and typically has the best survival outcome.

Selective estrogen receptor modulators, or SERMs, act directly on the estrogen receptor but in a way that allows it to be selective with the types of tissues it targets, therefore not completely shutting down ER activity [106]. Tamoxifen is a SERM and acts as an estrogen antagonist by binding to the same ligand binding region as estrogen, therefore blocking further estrogen binding [104]; however, transcription is not activated because there is an incomplete conformational change that occurs upon binding. Raloxifen is another SERM similar to Tamoxifen and can be further beneficial in preventative treatment because of a lower risk than Tamoxifen of causing endometrial carcinomas and complications [106]. Fulvestrant, a pure anti-estrogen, binds ER in a similar way to Tamoxifen; however it does not do so in a tissue selective manner, effectively shutting down ER activity completely [106]. Because of the full body nature for which Fulvestrant functions, it is typically used only once a resistance has been developed against Tamoxifen [106]. Resistance to Tamoxifen is common, 40% of ER alpha-positive tumors, with patients receiving endocrine therapy, and is the main cause when this therapy becomes ineffective [106, 108]. Strangely enough, 80% of hormone therapy-resistant mammary tumors will still have the same expression levels of estrogen receptors as they did prior to hormone treatment [105]. At this point, aromatase inhibitors can be started to try and treat the resistant tumors [106]. By blocking aromatization of androgens, you are blocking the supply of estrogen and indirectly blocking the function of estrogen receptors [106]. However, when a mammary carcinoma lacks an overexpression of estrogen receptors, progesterone receptors, and HER2, endocrine therapies will be ineffective because the proteins the drugs interact with are not present. The so-called triple-negative breast cancer (very similar in characteristics to basal-like carcinoma) has a very poor prognosis due to the inability to receive effective hormone therapies. Around 12–17% of women diagnosed with breast cancer are triple-negative [101]. For this group of carcinoma type, treatment options include chemotherapy in addition to platinum salts, PARP inhibitors, or anti-angiogenic agents [101].

The presence of estrogen and progesterone receptors in the tumor tissue correlates well with response to hormone therapy and chemotherapy [110, 111]. It is presently accepted the immune-histochemical method that has the versatility to be performed in minute amount of fixed tissue. The two parameters evaluated in immune-histochemical preparations of hormone receptors are the number of tumor cell nuclei stained and the intensity of the reaction. The first is expressed as a percentage of the entire tumor cell nuclei population, and the second is graded as negative, weak, moderate, and strong. The two parameters are sometimes combined into a scoring system, of which three major versions exist [112, 113]. The receptors are expressed in both ductal and lobular type of carcinomas; however medullary carcinomas and intraductal carcinomas of the comedocarcinoma type are negative, whereas mucinous carcinomas have the highest rates of positivity.

The Allred scoring system [112, 114] is used for evaluation of the presence of hormone receptor in the human breast cancer. Allred scoring has become the standard in the immune-histochemical analysis of mammary carcinomas and reduces the borderline or weak positive groups significantly. The grading system looks at the intensity of the positive stained cells and the proportion of the cells that are positive. Each of these characteristics receives a score, which then makes up the total score. The total score will be between 0 and 8, 8 being the highest positive scoring. A total score, greater than 2, results in a positive overall scoring for the tumor sample. Applying a combined endocrine receptor (CER) [123] has shown to be a more powerful predictor of 5-year DFS and BCSS than either ER or PR alone and resulted to be a more powerful discriminator of patient outcome than either ER or PR alone.

1.4.4 Role of HER2 in Breast Cancer Diagnosis and Treatment

Research into the nature of breast cancer has led to understanding the importance of certain genes and proteins, specifically how they affect the development and attribute unique features to the cancers. One of the most influential genes is the ERBB2 gene, also known as human epidermal growth factor receptor 2 (HER2), which codes for a protein called HER2 [115]. The ERBB2 gene itself was first identified in rats and humans in the early 1980s. The rat gene (identified first) was referred to as the *neu* oncogene after the transforming, cancer-producing gene found in a cell line of rat neural tumors [116]. It was found to be homologous to the avian erythroblastosis virus, or v-ERBB2 [117]. Thus, it was given the full name of v-erb-b2 avian erythroblastic leukemia viral oncogene homolog 2, symbolized by ERBB2 [118]. The human homologue gene was called HER2 as it related to the epidermal growth factor receptor (EGFR, aka ERBB1), forming a family of receptors [119]. Subsequent studies identified the two other members of the ERBB family: ERBB3 and ERBB4, also called HER3 and HER4, respectively [120, 121]. HER2 is a cell-surface receptor and a member of the EGFR family, consisting of structurally similar receptor-tyrosine kinases (RTKs, which are specifically activated by phosphorylating their tyrosine amino acid) [122, 123]. It contains three structural components characteristic to its EGFR family: an extracellular ligand-binding domain, a transmembrane domain, and an intracellular domain residing within the cytoplasm of the cell [124]. Although the HER2 receptors cannot directly bind to growth factors, they are able to dimerize with other EGFR proteins to function, forming heterodimers [125]. This is due to the extracellular domain, which is further divided into four sub-domains: Sub-domain I and III (bind together to block ligand access and enable dimerization), Sub-domain II (dimerize with the other receptors), and Sub-domain IV (believed to stabilize the receptor and open-conformation) [126, 127]. The entire process involves the other receptors becoming activated via a growth factor ligand, resulting in their conformational change needed to form dimers on the extracellular domains [127]. Upon forming the dimer, the intracellular domains reposition, allowing for exposure of several specific amino acids to be phosphorylated and adaptor proteins to be attached to a carboxyl-terminal tail of the domain [128]. In addition to this method, HER2 receptors are also able to autophosphorylate for ligand- independent signaling [129]. The general function of HER2 receptor, once dimerized, is to activate pathways leading to cellular proliferation and growth [130]. Of these heterodimers, the most active and potent is the HER2/HER3 dimer [131]. Among the pathways activated by HER2 are those of the mitogen-activated protein kinase (MAPK). A MAPK group known as extracellular signal-regulated kinases (ERKs), upon activation, is able to increase their activity levels, relocate to the cellular nucleus to alter transcription and gene expression, and ultimately promote mitosis and differentiation while inhibiting apoptosis [132, 133]. The Phosphoinositide 3-Kinase/Protein Kinase B (PI3K/Akt) pathway displays similar functions, where activated PI3K catalyzes signaling molecules leading to Akt activation, which in turn inhibits apoptosis and promoting cell growth [134]. As HER2 is a proto-oncogene, overexpression or amplification of HER2 thus leads to uncontrolled cellular proliferation, known to lead to mammary tumorigenesis, aggressive tumor growth (due to angiogenesis), and even metastasis [135–137].

Clinically, HER2 expression breast cancers occur in 20–30% of all breast cancer cases [138]. The most common carcinomas occur within the milk ducts of the breast, known as ductal carcinoma, while others occur within the milk-secreting glands, or lobules, of the breast, known as lobular carcinoma [139]. These carcinomas can be identified by their cellular proliferation invading into other organs and tissues as invasive carcinomas, but can also remain situated within the original tissue as in situ carcinomas [140]. HER2 amplification has been found mostly in about 40% of ductal carcinomas [141]. Lobular carcinomas rarely express HER2, and often only in the pleomorphic variant [142].

Several different methods utilize HER2 overexpression and amplification to obtain diagnostic information. Some of the most commonly used are immunohistochemistry (IHC) tests, which utilize specific antibodies to detect antigens in the cells of a tissue section, allowing for diagnostics of abnormal cells in cancerous tumors [143]. Thus, expression levels of HER2 can be detected on the cell surfaces. HER2 is expressed in both HER2/neu+ and luminal B (HER2+) breast cancers. These subtypes are generally regarded to be more aggressive (actively spread into other organs and systems), have poorer tumor grades (poorly differentiated cells), affect the lymph nodes, have higher recurrence rates, and have an overall poorer prognosis than the other subtypes [144, 145]. Despite the poor prognosis associated with these breast cancers, several treatments are available, including several drugs specific to the HER2/neu receptor. These drugs include Trastuzumab (commercially known as Herceptin) and Pertuzumab. Both are monoclonal antibodies, or antibodies produced by cloned immune cells from a single parent cell, specifically cancer cells overexpressing the HER2/neu receptor [146]. Trastuzumab antibodies are able to readily bind to the HER2/neu receptor on the extracellular component, inhibiting the intracellular domain to down-regulate the downstream pathways, and ultimately reduce proliferation of the breast cancer cells [147]. Pertuzumab functions in the same manner, but instead targets the HER2 receptor directly to prevent dimerization with other receptors [148].

Trastuzumab itself is known, in rare cases, to cause cardiac disorders such as cardiomyopathy [149]. Other treatments less specific to HER2 have also shown to be effective. This includes chemotherapeutic regiments utilizing an anthracycline agent, such as doxorubicin, and a combination of other agents [150]. Unfortunately, hormone therapies do not seem to be effective against HER2+.

In summary, HER2/neu (c-erbB-2) is an oncogene that encodes a transmembrane glycoprotein with tyrosine kinase activity known as p185, which belongs to the family of epidermal growth factor receptors [151, 152]. Its overexpression can be measured by immunohistochemistry or FISH [153], and a good correlation exists between these methods [154–156]. When the results are either 3+ or 0 by immunohistochemistry, no further validation is needed, but if the immunotest gives instead a result of 1+ or 2+, the performance of FISH is recommended. Overexpression of HER2/neu by either technique is a very good predictor of response to Herceptin, but not a very good predictor of response to chemotherapy or overall survival.

1.4.5 Evaluation of Ki-67

Ki-67 is a protein found in the nucleus of the cell. The protein was defined by an antibody that was generated by the nuclei of the Hodgkin lymphoma cell line in immunized mice. The antigen was later discovered to be a protein, but the name Ki-67, after the location of discovery (Kiel University, Germany), and the clone number in the 96-well plate remained the same [157, 158]. Ki-67 is expressed in the G1, S, and G2 phases of cell division and mitosis, but not during the G0 phase, also known as the resting phase [158]. The expression levels change throughout the different phases, with low expression in the G1 and S phases, the highest expression during mitosis [157]. This is followed by a decrease in expression during the anaphase and telophase at the end of mitosis [157]. Due to the presence of Ki-67 in all human cell types during the cell cycle and its quick disappearance from the cell after the end of the proliferative state, it is used as a marker for cell proliferation [156]. While much is known about Ki-67's location in the cell, its expression in relation to cell proliferation, and its primary structure, little is known about its actual function in the cell [158]. The usual methods of finding the most effective treatment for breast cancer are determining the grade, stage, and endocrine status of the tumor. While not required, it is recommended that a proliferation marker such as Ki-67 is used. It also has predictive and prognostic applications. Ki-67 levels are low in healthy breast tissue. In the case of ductal carcinoma in situ, high expression of Ki-67 is associated with a negative prognosis and higher chance for recurrence [159]. Similarly, in triple-negative breast cancer, a high Ki-67 index is an indicator for a good response to chemotherapy, but a poor overall prognosis [159]. Ki-67 levels are used to indicate the chance of tumor recurrence in relation to time. Tumor types with higher Ki-67 expression, such as HER-2 enriched tumors and triple-negative tumors, have a significantly higher initial risk of recurrence. In the luminal A and luminal B subtypes, the expression of Ki-67 is low, and therefore the risk of recurrence is much lower [160].

Another protein called separase has been shown to play a role in cell cycle progression by unleashing sister-chromatids at anaphase onset. This protein has been proposed to be used as another marker of cell proliferation and has been shown that its expression has a strong and independent prognostic value [161].

1.4.6 Tests for Molecular Profiling of Breast Cancer

After the publication of the new genomic classification of breast cancer [162], several adaptations compiling few representative genes have been introduced in the practice of oncology. Among them is the Oncotype Dx that contains 21 gene expression signature, and the first major trial was published in 2004 [163]. This study comprises women that have ER+ and LN-negative breast cancer. This test was recommended by NCCN and ASCO. The Oncotype was developed from a FFPE

(Formalin-Fixed Paraffin-Embedded) compliant assay to predict distant recurrence of ER+ breast cancer and originally selected 250 candidate genes to test on NSABP B-14 and B-20 trials. At the end they refined a 16+5 gene panel that could predict recurrence [163].

The MammaPrint contains 70 gene expression signature, and the major trial was from 2002 comprising women <61 years, T1-T2, and N0 disease [164]. The Prosigna from NanoString contains 50 gene expression signature +5 control genes and represents the old PAM50 assay. The PAM or prediction analysis of microarrays recapitulated the microarray classifier using RT-PCR-based PAM50 assay in comparison to standard clinical molecular markers. The major trial was in 2013 using stage I–III cancer population and cleared by the FDA in 2013. Thirteen years after Perou's paper [162], the PAM50 is entering the clinical arena. The PAM50 gene signature has been transferred to a novel and robust method for mRNA quantification [165]. The methods work well in FFPE, does not rely on amplification of nucleic acids, and is intended for kit use in local labs with the proper instruments. The PAM50 expression results are used to calculate a risk of recurrence score (ROR) and provide low-, intermediate-, and high-risk groups. The score is based on the intrinsic subtype and pathologic characteristics (T,N), with special weighting given to a set of proliferation associated genes. PAM50 correlates well with the Oncotype and the use of the four IIC parameters (ER, PR, HER2, and Ki-67). Table 1.4 summarized the major features of these three genomic tests in breast cancer.

1.4.7 General Considerations

For the prognosis of breast cancer, oncologists find it helpful to know the tumor grade as well as the size of the tumor and involvement of the lymph nodes, and there is no doubt that the practical oncologist want something more than ER, PR, and HER2 results such Ki-67, a nuclear protein associated with cellular proliferation;

Table 1.4 Major features of three genomic tests in breast cancer

	MammaPrint	Oncotype Dx	Prosigna
Input material	Fresh frozen FFPE	FFPE	FFPE
Platform	Microarray	qPCR	nCounter
#genes analyzed	70	21	50+5
Target patient population	Stage I-II	ER+ Stage I-II	Stage I-II (Stage 1-III)
Regulatory	FDA cleared (frozen)	NCCN/ASCO Guidelines	FDA cleared
Performance site	Central	Central	Decentralized kit format
Features	Binary stratification; molecular subtypes; lots of data	Gold standard; ER+ cancer; now in DCIS	ROR compares well; innovative technology; *intrinsic subtypes*

PI3K (phosphatidylinositol 3-kinase) mutations; and tumor-infiltrating lymphocytes and how to use next-generation sequencing panels.

ER, PR, and HER2 have established clinical value and are used widely and with confidence. However, PR is considered minor when compared with ER that determines most of the time the receptivity for hormone therapy. It is well known that tumors that are ER and PR positive do better than those that are ER positive/PR negative, but PR alone may be sufficient to confer some sensitivity to anti-estrogen therapy and those cases that are PR positive but ER negative remain candidates for endocrine anti-estrogen therapy. Therefore, the evaluation of PR cannot be underestimated. The biological basis of this is that PR gene is regulated by ER and usually is detected in tumor cells with an activated ER signaling pathway. An important problem that needs to be acknowledged is that there are no good antibodies for PR like the ones that are offered for ER. The values of ER status or hormonal status of the tumors are important in view that some patients with ER-positive tumors benefit from extended hormonal therapy for 10 years instead of the 5 years previously recommended.

Whereas HER2-positive tumors are managed by the oncologist, the resistance of those tumors to the treatment with HER2-targeted drugs has pushed to develop the T-DM1, also known as trastuzumab emtansine, a conjugate of Herceptin and a chemotherapeutic agent that appears to be remarkably effective in patients whose tumors are HER2 positive and who have become resistant to Herceptin.

A problem that oncologists and pathologist are facing is the evaluation of Ki-67. High Ki-67 is linked with high risk of recurrence in ER-positive cancers and is associated with greater responsiveness to chemotherapy. However, the evaluation of Ki-67 is not well standardized with large discordance in how labs assign Ki-67 results. The new generation of prognostic markers that are commercially available are Prosigna (NanoString Technologies), Breast Cancer Index (bioTheranostics), Oncotype DX (Genomic Health), and MammaPrint (Agendia) and already provide prognostic information; however they are not widely used, and these next-generation tests are not helping the oncologist to make the right decisions about which drug or specific therapy to choose. Even the problem is more severe with the sequencing based tests will likely be most useful in cases involving patients with metastatic disease and here the problem is how to treat a mutation and is the right drug for the right target available or ready to be used?

In this chapter we wanted that the reader be aware that still we do not have the ultimate prognostic and predictive test and in some cases the right interpretation of the available test is confusing for the oncologist, the pathologist, and the patient.

1.5 The Present and the Future Direction of the Genomic Signature of Breast Cancer

The emergence of better and faster DNA sequence analysis has allowed to characterize the mutations of known as well as less explored parts of the genome such as base substitutions, small insertions/deletions, rearrangements, and copy number changes [166–170]. This new set of data [171] has provided a better understanding of the role of activating driver rearrangements [172–174], forming chimeric genes/proteins or relocating genes adjacent to new regulatory regions, and also a better understanding of the role of driver substitutions and indels in non-coding regions of the genome [175, 176]. Indel is a term coined for identifying an *in*sertion or deletion of bases in the genome of an organism. It is classified among small genetic variations, measuring from 1 to 10,000 base pairs in length and different from a microindel, that is, an indel that results in a net change of 1–50 nucleotides. These techniques have also allowed to identify which mutational processes generate the somatic mutations found in breast cancer [177, 178].

In the study of the whole genomes of 560 breast cancers and non-neoplastic tissue from each individual [171] were sequenced and detected 3,479,652 somatic base substitutions, 371,993 small indels, and 77,695 rearrangements, with substantial variation in the number of each between individual samples. Five cancer genes (*MED23*, *FOXP1*, *MLLT4*, *XBP1*, *ZFP36L1*) were found for which evidence was previously absent or equivocal, or for which the mutations indicate the gene acts in breast cancer in a recessive rather than in a dominant fashion, as previously reported in other cancer types [173]. The ten most frequently mutated genes were *TP53*, *PIK3CA*, *MYC*, *CCND1*, *PTEN*, *ERBB2*, *ZNF703/FGFR1* locus, *GATA3*, *RB1*, and *MAP 3K1. These genes represented* 62% of drivers. Driver mutations are those that confer proliferative advantage on a cell, leading to outgrowth of a neoplastic clone [179]. Most of these driver mutations arise in somatic cells during the lifetime of the cancer patient at difference of the "passenger" mutations that are not directly implicated in cancer development [179].

Mutational processes generating somatic mutations imprint particular patterns of mutations on cancer genomes, and in the 560 breast cancer sequenced [171], 12 signatures, including those previously observed, and a further seven, of which five have formerly been detected in other cancer types and two new were found. These signatures were further sub-classified into 32 subclasses. Of interest was the finding rearrangement signature 3 in which 95% of rearrangement in this signature were concentrated in 15% of cancers and 91% of the cancers with *BRCA1* mutations or promoter hypermethylation were in this group, which was enriched for basal-like, triple-negative cancers and copy number classification of a high homologous recombination deficiency (HRD) index 38–40. Thus, inactivation of *BRCA1*, but not *BRCA2*, may be responsible for the rearrangement signature 3 small tandem duplication mutator phenotype [171]. Altogether *this analysis of all classes of somatic mutation across exons, introns, and intergenic regions highlights the repertoire of cancer genes and mutational processes operating, and progresses towards a comprehensive account of the somatic genetic basis of breast cancer, and how all this vast information will allow to identify tools for treatment and prevention is part of the future of breast cancer research.*

1.6 Future of the Use of *Liquid Biopsy*

At difference of tissue biopsy that provides an information frozen in time, the liquid biopsy encompasses assays that measure tumor-derived cells or products in blood and other body fluids that, in general, may be obtained serially and easily with minimal patient risk. Most commonly studied are circulating tumor cells or CTCs circulating tumor DNA (ctDNA), microRNAs [180], analyses of extracellular vesicles, including exosomes, microvesicles, large oncosomes, and apoptotic bodies [181–188], immune-derived cells [189–191], tumor-derived circulating endothelial cells [192, 193] and tumor-educated platelets [194–196]. The advantage of liquid biopsy is that not only could it be used for diagnosis but also for following the course of the disease and for assessment of cancer surveillance and monitoring for relapse and treatment response. Eventually this technology will be able to screen asymptomatic individuals for early detection of cancer and for evaluating preventative strategies for breast cancer prevention and control. There are two programs underway like Grail [197] and STRIVE study [198] that will provide significant information in thousands of human samples, and it will be an invaluable resource in the use of liquid biopsy.

References Cited

1. Fu, J., Wu, L., Fu, W., Tan, Y., Xu, T., Hong, Z., Wang, F., & Li, S. (2018). How young is too young in breast cancer?-Young breast cancer is not a unique biological subtype. *Clinical Breast Cancer, 18*, e25–e39.
2. Aamdal, S., et al. (1984). Estrogen receptor and long-term prognosis in breast cancer. *Cancer, 53*, 2525–2529.
3. Black, M. M., & Speer, F. D. (1957). Nuclear' structure in cancer tissues. *Surgery, Gynecology & Obstetrics, 105*, 97–102.
4. Bloom, H. J. G., & Richardson, W. W. (1957). Histologic grading and prognosis in breast cancer: A study of 1409 cases of which 359 have been followed for 15 years. *British Journal of Cancer, 11*, 359–377.
5. Fisher, B., & Slack, N. H. (1970). Number of lymph nodes examined and the prognosis of breast cancer. *Surgery, Gynecology & Obstetrics, 131*, 79–88.
6. Fisher, E. R., Gregorio, R. M., & Fisher, B. (1975). The pathology of invasive breast cancer-a syllabus derived from findings of the National Surgical Adjuvant Breast Project (Protocol N. 4). *Cancer, 36*, 1–84.
7. Fisher, E. R., Palekar, A. S., Redmond, C., Barton, N., & Fisher, B. (1980). Pathologic findings from the National Surgical Adjuvant Breast Project. (Protocol N. 4). VI. Invasive papillary cancer. *American Journal of Clinical Pathology, 73*, 313–322.
8. Fisher, E. R., Redmond, C. K., Lieu, H., Rockette, H., & Fisher, B. (1980). Collaborating NSABP Investigators: correlation of estrogen receptor and pathologic characteristics of invasive breast cancer. *Cancer, 45*, 349–353.
9. Foote, F. W., Jr. (1959). Surgical pathology of cancer of the breast. In W. H. Parsons (Ed.), *Cancer of the breast* (pp. 37–38). Springfield: Charles C Thomas.

10. Knight, W. A., Livingston, R. B., Bregory, E. J., & McGuire, W. L. (1977). Estrogen receptor as an independent prognostic factor for early recurrence in breast cancer. *Cancer Research, 37*, 4669–4671.
11. Lane, N., Gokiel, H., Salerno, R. A., & Haagensen, C. D. (1961). Clinic pathologic analysis of the surgical curability of breast cancers: A minimum ten-year study of a personal series. *The American Surgeon, 153*, 483–498.
12. McDivitt, R. W., Stewart, F. W., & Berg, J. W. (1968). *Tumors of the breast, Atlas of tumor pathology, second series*. Washington, DC: Armed Forces Institute of Pathology.
13. McGuire, W. L. (1975). Current status of estrogen receptors in human breast cancer. *Cancer, 36*, 638–644.
14. Rich, M. A., Furmanski, P., & Brooks, S. C. (1978). The breast cancer prognostic study surgery and pathology associates: Prognostic value of estrogen receptor. *Cancer Research, 38*, 4296–4298.
15. Rosen, P. P. (1979). The pathological classification of human mammary carcinoma, past, present and future. *Annals of Clinical and Laboratory Science, 144*, 156.
16. Russo, J. (2016). *The pathobiology of breast cancer*. New York: Springer.
17. Rosai, J. (2004). Breast. Chapter 20. In *Rosai and Ackerman's surgical pathology* (pp. 1802–1818). New York: Mosby.
18. Kuhajda, F. P., Offutt, L. E., & Mendelsohn, G. (1983). The distribution of carcinoembryonic antigen in breast carcinoma. Diagnostic and prognostic implications. *Cancer, 52*, 1257–1264.
19. Loy, T. S., Chapman, R. K., Diaz-Arias, A. A., Bulatao, I. S., & Bickel, J. T. (1991). Distribution of BCA-225 in adenocarcinomas. An immunohistochemical study of 446 cases. *American Journal of Clinical Pathology, 96*, 326–329.
20. Prey, M. U., Bedrossian, C. W., & Masood, S. (1991). The value of monoclonal antibody B72.3 for the diagnosis of breast carcinoma. Experience with the first commercially available source. *Human Pathology, 22*, 598–602.
21. Tavassoli, F. A., Jones, M. W., Majeste, R. M., Bratthauer, G. L., & O'Leary, T. J. (1990). Immunohistochemical staining with monoclonal Ab B72.3 in benign and malignant breast disease. *The American Journal of Surgical Pathology, 14*, 128–133.
22. Guerreiro Da Silva, I. D., Hu, Y. F., Russo, I. H., Ao, X., Salicioni, A. M., Yang, X., & Russo, J. (2000). S100P Ca +2 -binding protein overexpression is associated with immortalization and neoplastic transformation of human breast epithelial cells *in vitro* and tumor progression *in vivo*. *International Journal of Oncology, 16*, 231–240.
23. Dwarakanath, S., Lee, A. K. C., DeLellis, R. A., Silverman, M. L., Frasca, L., & Wolfe, H. I. (1987). S-100 protein positivity in breast carcinomas. A potential pitfall in diagnostic immunohistochemistry. *Human Pathology, 18*, 1144–1148.
24. Lunde, S., Nesland, J. M., Holm, R., & Johannessen, J. V. (1987). Breast carcinomas with protein S-100 immunoreactivity. An immunocytochemical and ultrastructural study. *Pathology, Research and Practice, 182*, 627–631.
25. Huang, Y., Fernandez, S., Goodwin, S., Russo, P. A., Russo, I. H., Sutter, T., & Russo, J. (2007). Epithelial to mesenchymal transition in human breast epithelial cells transformed by 17- beta- estradiol. *Cancer Research, 67*, 11147–11157.
26. Russo, J., Tait, L., & Russo, I. H. (1986). Guidelines in the ultrastructural diagnosis of tumors. In J. Russo & S. C. Somers (Eds.), *Tumor diagnosis by electron microscopy* (pp. 15–126). New York: Field, Rich and Associates.
27. Roxanis, I., Colling, R., Kartsonaki, C., Green, A. R., & Rakha, E. A. (2018). The significance of tumor micro architectural features in breast cancer prognosis: A digital image analysis. *Breast Cancer Research, 20*, 11.
28. Page, D. L., Dixon, J. M., Anderson, T. J., Lee, D., & Stewart, H. J. (1983). Invasive cribriform carcinoma of the breast. *Histopathology, 7*, 525–536.
29. Venable, J. G., Schwartz, A. M., & Silverberg, S. G. (1990). Infiltrating cribriform carcinoma of the breast. A distinctive clinic pathologic entity. *Human Pathology, 21*, 333–338.

30. Walker, R. A. (1982). Mucoid carcinomas of the breast. A study using mucin histochemistry and peanut lectin. *Histopathology, 6*, 571–579.
31. Sáez, C., Japón, M. A., Poveda, M. A., & Segura, D. I. (2001). Mucinous (colloid) adenocarcinomas secrete distinct O-acylated forms of sialomucins: A histochemical study of gastric, colorectal and breast adenocarcinomas. *Histopathology, 39*, 554–560.
32. O'Connell, J. T., Shao, Z. M., Drori, E., Basbaum, C. B., & Barsky, S. H. (1998). Altered mucin expression is a field change that accompanies mucinous (colloid) breast carcinoma histogenesis. *Human Pathology, 29*, 1517–1523.
33. Norris, H. I., & Taylor, H. E. (1965). Prognosis of mucinous (gelatinous) carcinoma of the breast. *Cancer, 18*, 879–885.
34. Rasmussen, B. B., Rose, C., & Christensen, I. B. (1987). Prognostic factors in primary mucinous breast carcinoma. *American Journal of Clinical Pathology, 87*, 155–160.
35. McDivitt, R. W., Boyce, W., & Gersell, D. (1982). Tubular carcinoma of the breast. Clinical and pathological observations concerning 135 cases. *The American Journal of Surgical Pathology, 6*, 401–411.
36. Peters, G. N., Wolff, Y. I., & Haagensen, C. D. (1981). Tubular carcinoma of the breast. Clinical pathologic correlations based on 100 cases. *Annals of Surgery, 193*, 138–149.
37. Carstens, P. H. B., Huvos, A. G., Foote, F. I. N., Jr., & Ashikari, R. (1972). Tubular carcinoma of the breast. A clinic pathologic study of 35 cases. *American Journal of Clinical Pathology, 58*, 231–238.
38. Deos, P. H., & Noris, H. J. (1982). Well-differentiated (tubular) carcinoma of the breast. A clinicopathologic study of 145 pure and mixed cases. *American Journal of Clinical Pathology, 78*, 1–7.
39. Diab, S. G., Clark, G. M., Osborne, C. K., Libby, A., Allred, D. C., & Elledge, R. M. (1999). Tumor characteristics and clinical outcome of tubular and mucinous breast carcinomas. *Journal of Clinical Oncology, 17*, 1442–1448.
40. Bloom, H. J. G., Richardson, W. W., & Fields, J. R. (1970). Host resistance and survival in carcinoma of breasts. A study of 104 cases of medullary carcinoma in a series of 1,411 cases of breast cancer followed for 20 years. *British Medical Journal, 3*, 181–188.
41. Richardson, W. W. (1956). Medullary carcinoma of the breast. A distinctive tumor type with a relatively good prognosis following radical mastectomy. *British Journal of Cancer, 10*, 415–423.
42. Abati, A. D., Kimmel, M., & Rosen, P. P. (1990). Apocrine mammary carcinoma. A clinic pathologic study of 72 cases. *American Journal of Clinical Pathology, 94*, 371–377.
43. McDivitt, R. W., & Stewart, F. W. (1966). Breast carcinoma in children. *Journal of the American Medical Association, 195*, 388–390.
44. Rosen, P. P., & Cranor, M. L. (1991). Secretory carcinoma of the breast. *Archives of Pathology & Laboratory Medicine, 115*, 141–144.
45. Tavassoli, F. A., & Norris, H. J. (1980). Secretory carcinoma of the breast. *Cancer, 45*, 2404–2413.
46. Bussolati, G., Papotti, M., Sapino, A., Gugliotta, P., Ghiringhello, B., & Azzopard, J. G. (1987). Endocrine markers in argyrophilic carcinomas of the breast. *The American Journal of Surgical Pathology, 11*, 248–256.
47. Oberman, H. A. (1987). Metaplastic carcinoma of the breast. A clinicopathologic study of 29 patients. *The American Journal of Surgical Pathology, 11*, 918–929.
48. Acs, G., Lawton, T. J., Rebbeck, T. R., Li Volsi, V. A., & Zhang, P. J. (2001). Differential expression of E-cadherin in lobular ductal neoplasms of the breast and its biologic and diagnostic implications. *American Journal of Clinical Pathology, 115*, 85–98.
49. Domagala, W., Harezga, B., Szadowska, A., Markiewski, M., Weber, K., & Osborn, M. (1993). Nuclear protein accumulates preferentially in medullary and high grade ductal but in lobular breast carcinomas. *The American Journal of Pathology, 142*, 669–674.
50. Goldstein, N. S., Bassi, D., Watts, J. C., Layfield, L. J., Yaziji, H., & Gown, A. M. (2001). E-cadherin reactivity of 95 non-invasive ductal and lobular lesions of the breast implications

for the interpretation of problematic lesions. *American Journal of Clinical Pathology, 115*, 534–542.
51. Lehr, H. A., Folpe, A., Yaziji, H., Kommoss, F., & Gown, A. M. (2000). Cytokeratin 8 immunostaining pattern and E-cadherin expression distinguish lobular from ductal breast carcinoma. *American Journal of Clinical Pathology, 114*, 190–196.
52. Weidner, N., & Semple, J. P. (1992). Pleomorphic variant of invasive lobular carcinoma of the breast. *Human Pathology, 23*, 1167–1171.
53. Frolik, D., Caduff, R., & Varga, Z. (2001). Pleomorphic lobular carcinoma of the breast: Its cell kinetics, of oncogenes and tumor suppressor genes compared with invasive ductal carcinomas and classification infiltrating lobular carcinomas. *Histopathology, 39*, 503–513.
54. Middleton, L. P., Palacios, D. M., Bryant, B. R., Krebs, P., Otis, C. N., & Merino, M. J. (2000). Pleomorphic lobular carcinoma: Morphology, immunohistochemistry, and molecular analysis. *The American Journal of Surgical Pathology, 24*, 1650–1656.
55. Radhi, J. M. (2000). Immunohistochemical analysis of pleomorphic lobular carcinoma: Higher expression of p53 and chromogranin and lower expression of ER and PgR. *Histopathology, 36*, 156–160.
56. Wahed, A., Connelly, J., & Reese, T. (2000). E-cadherin expression in pleomorphic lobular carcinoma: An aid to differentiation from ductal carcinoma. *Annals of Diagnostic Pathology, 26*, 349–351.
57. Fisher, E. R., Gregorio, R. M., Redmond, C. I., & Fisher, B. (1977). Tubulolobular invasive breast cancer: A variant of lobular invasive cancer. *Human Pathology, 8*, 679–683.
58. Padmore, R. F., Fowble, B., Hoffman, J., Rosser, C., Hanlon, A., & Patchefsky, A. S. (2000). Microinvasive breast carcinoma; clinic pathologic analysis of a single institution experience. *Cancer, 88*, 1403–1409.
59. Paget, J. (1874). On disease of the mammary areola preceding cancer of the mammary gland. *St Barth Hospital Reproductive, 10*, 87–89.
60. Perou, C. M., Sorlie, T., et al. (2000). Molecular portraits of human breast tumors. *Nature, 406*, 747–752.
61. Sorlie, T., Perou, M., et al. (2001). Gene expression patterns of breast carcinomas distinguish tumor subclasses with clinical implications. *Proceedings of the National Academy of Sciences of the United States of America, 98*, 10869–10874.
62. Nishimura, R., et al. (2010). Ki-67 as a prognostic marker according to breast cancer subtype and a predictor of recurrence time in primary breast cancer. *Experimental and Therapeutic Medicine, 1*, 747–754.
63. Goldhirsch, A., et al. (2013). Strategies for subtypes—Dealing with the diversity of breast cancer: Highlights of the St Gallen International Expert Consensus on the Primary Therapy of Early Breast Cancer 2011. *Annals of Oncology, 22*, 1736–1747.
64. Nielsen, T. O., Hsu, F. D., Jensen, K., Cheang, M., Karaca, G., Hu, Z., et al. (2004). Immunohistochemical and clinical characterization of the basal-like subtype of invasive breast carcinoma. *Clinical Cancer Research, 10*, 5367–5374.
65. Prat, A., & Perou, C. M. (2011). Deconstructing the molecular portraits of breast cancer. *Molecular Oncology, 5*, 5–23.
66. Gao, J. J., & Swain, S. M. (2018). Luminal a breast cancer and molecular assays: A review. *The Oncologist, 10*, 1634.
67. Dawson, S. J., Rueda, O. M., et al. (2013). A new genome-driven integrated classification of breast cancer and its implications. *The EMBO Journal, 32*, 617–628.
68. Santagata, S., Thakkar, A., et al. (2014). Taxonomy of breast cancer based on normal cell phenotype predicts outcome. *The Journal of Clinical Investigation, 124*, 859–870.
69. Swerdlow, S. H. (2013). Lymphoma classification and the tools of our trade: An introduction to the 2012 USCAP Long Course. *Modern Pathology, 26*(Suppl 1), S1–S14.
70. Antomiades, K., & Spector, H. (1979). Correlation of estrogen receptor levels with histology and cytomorphology in human mammary cancer. *American Journal of Clinical Pathology, 71*, 497–503.

71. Bishop, H. M., Blarney, R. W., Elston, C. W., Hoybittle, J. L., Nicholson, R. I., & Griffiths, K. (1979). Relationship of estrogen-receptor status to survival in breast cancer. *Lancet, 1*, 283–284.
72. Black, M. M., Opler, S. R., & Speer, F. D. (1955). Survival in breast cancer cases in relation to the structure of the primary tumor and regional lymph nodes. *Surgery, Gynecology & Obstetrics, 100*, 543–551.
73. Black, M. M., Speer, F. D., & Opler, S. R. (1956). Structural representations of tumor-host relationships in mammary carcinoma. Biologic prognostic significance. *American Journal of Clinical Pathology, 26*, 250–265.
74. Bloom, H. J. G., Richardson, W. W., & Field, J. R. (1970). Host resistance and survival in carcinoma of breast. Study of 104 cases of medullary carcinoma in a series of 1411 cases of breast cancer followed for 20 years. *British Medical Journal, 3*, 181–188.
75. Cutler, S. J., Black, M. M., & Goldenberg, I. S. (1963). Prognostic factor in cancer of the female breast. *Cancer, 16*, 1589–1597.
76. Cutler, S. J., Black, M. M., Friedell, G. H., Vidone, R. A., & Goldenberg, I. S. (1966). Prognostic factors in cancer of the female breast. II, Reproducibility of histopathologic classification. *Cancer, 19*, 75–82.
77. Fisher, B., Glass, A., Redmond, C., et al. (1977). L-Phenylalanine mustard (L-PAM) in the management of primary breast cancer: An update of earlier findings and a comparison of the utilizing L.PAM plus 5 fluorouracil (5FU). *Cancer, 39*, 2883–2903.
78. Fisher, E. R., Gregorio, R. M., & Fisher, B. (1974). The pathology of invasive breast cancer-a syllabus derived from findings of the National Surgical Adjuvant Breast Project (Protocol N. 4). *Cancer, 36*, 1–84.
79. Fisher, E. R., Redmond, C. K., Lieu, H., Rockette, H., & Fisher, B. (1980). Collaborating NSABP Investigators: Correlation of estrogen receptor and pathologic characteristics of invasive breast cancer. *Cancer, 45*, 349–353.
80. Hahnel, R., Wooding, T., & Vivian, A. B. (1979). Prognostic value of estrogen receptor in primary breast cancer. *Cancer, 44*, 671–675.
81. Lane, N., Gokiel, H., Salerno, R. A., & Haagensen, C. D. (1961). RECURRENCE AND SURVIVAL IN BREAST CANCER 131 gic analysis of the surgical curability of breast cancers: A minimum ten-year study of a personal series. *The American Surgeon, 153*, 483–498.
82. Masters, J. R. W., Sangster, K., Hawkins, R. A., & Shivas, A. A. (1976). Elastosis and estrogen receptors in human breast cancer. *British Journal of Cancer, 33*, 342–343.
83. Maynard, P. V., Blaney, R. N., Elston, C. W., Haybittle, J. L., & Griffiths, K. (1978). Estrogen receptor assay in primary breast cancer and early recurrence of the disease. *Cancer Research, 38*, 4292–4295.
84. Patchefsky, A., Shaber, G. S., Schwartz, G. F., Feig, S. A., & Norlinger, R. E. (1977). The pathology of breast cancer detected by mass population screening. *Cancer, 40*, 1659–1670.
85. Rosen, P. P., Lieberman, P. H., Braun, D. W., Jr., Koslolf, C., & Adair, F. (1978). Lobular carcinoma *in situ* of the breast. Detailed analysis of 99 patients with average follow up of 24 years. *The American Journal of Surgical Pathology, 2*, 225–251.
86. Russo, J., Frederick, J., Ownby, H. E., Fine, G., Husain, M., Krickstein, H. I., Robbins, T. O., & Rosenberg, B. F. (1987). Predictors of recurrence and survival of breast cancer patients. *American Journal of Clinical Pathology, 88*, 132–138.
87. Ohtani, H. (2007). Focus on TILs: Prognostic significance of tumor infiltrating lymphocytes in human colorectal cancer. *Cancer Immunity, 7*, 4.
88. Liakou, C. I., Narayanan, S., Tang, D. N., et al. (2007). Focus on TILs: Prognostic significance of tumor infiltrating lymphocytes in human bladder cancer. *Cancer Immunity, 7*, 10.
89. Uppaluri, R., Dunn, G. P., & Lewis, J. S., Jr. (2008). Focus on TILs: Prognostic significance of tumor infiltrating lymphocytes in head and neck cancers. *Cancer Immunity, 8*, 16.
90. Oble, D. A., Loewe, R., & Yu Pand Mihm, M. C., Jr. (2009). Focus on TILs: Prognostic significance of tumor infiltrating lymphocytes in human melanoma. *Cancer Immunity, 9*, 3.

91. Denkert, C., Loibl, S., Noske, A., et al. (2010). Tumor-associated lymphocytes as an independent predictor of response to neo-adjuvant chemotherapy in breast cancer. *Journal of Clinical Oncology, 28*, 105–113.
92. West, N. R., Milne, K., Truong, P. T., et al. (2011). Tumor-infiltrating lymphocytes predict response to anthracycline-based chemotherapy in estrogen receptor-negative breast cancer. *Breast Cancer Research, 13*, R126.
93. DeNardo, D. G., Brennan, D. J., Rexhepaj, E., et al. (2011). Leukocyte complexity predicts breast cancer survival and functionally regulates response to chemotherapy. *Cancer Discovery, 1*, 54–67.
94. Yamaguchi, R., Tanaka, M., Yano, A., et al. (2012). Tumor-infiltrating lymphocytes are important pathologic predictors for neoadjuvant chemotherapy in patients with breast cancer. *Human Pathology, 43*, 1688–1694.
95. Denkert, C., von Minckwitz, G., Brase, J. C., et al. (2015). Tumor-infiltrating lymphocytes and response to neoadjuvant chemotherapy with or without Carboplatin in human epidermal growth factor receptor 2-positive and triple-negative primary breast cancers. *Journal of Clinical Oncology, 33*, 983–991.
96. Kreike, B., van Kouwenhove, M., Horlings, H., et al. (2007). Gene expression profiling and histopathological characterization of triple-negative/basal-like breast carcinomas. *Breast Cancer Research, 9*, R65.
97. Loi, S., Sirtaine, N., Piette, F., et al. (2013). Prognostic and predictive value of tumor-infiltrating lymphocytes in a phase III randomized adjuvant breast cancer trial in node-positive breast cancer comparing the addition of docetaxel to doxorubicin with doxorubicin-based chemotherapy: BIG 02–98. *Journal of Clinical Oncology, 31*, 860–867.
98. Adams, S., Gray, R. J., Demaria, S., et al. (2014). Prognostic value of tumor-infiltrating lymphocytes in triple-negative breast cancers from two phase III randomized adjuvant breast cancer trials: ECOG 2197 and ECOG 1199. *Journal of Clinical Oncology, 32*, 2959–2966.
99. Salgado, R., Denkert, C., Demaria, S., et al. (2015). The evaluation of tumor-infiltrating lymphocytes (TILs) in breast cancer: Recommendations by an International TILs Working Group 2014. *Annals of Oncology, 26*, 259–271.
100. Yakirevich, E., Izhak, O. B., Rennert, G., et al. (1999). Cytotoxic phenotype of tumor infiltrating lymphocytes in medullary carcinoma of the breast. *Modern Pathology, 12*, 1050–1056.
101. Foulkes, W. D., Smith, I. E., & Reis-Filho, J. S. (2010). Triple-negative breast cancer. *The New England Journal of Medicine, 363*, 1938–1948.
102. Coates, A. S., Winer, E. P., Goldhirsch, A., et al. (2015). Tailoring therapies—Improving the management of early breast cancer: St Gallen International Expert Consensus on the Primary Therapy of Early Breast Cancer. *Annals of Oncology, 26*, 1533–1546.
103. Ohtani, H., Mori-Shiraishi, K., Nakajima, M., & Ueki, H. (2015). Defining lymphocyte-predominant breast cancer by the proportion of lymphocyte-rich stroma and its significance in routine histopathological diagnosis. *Pathology International, 65*, 644–651.
104. Nelson, L. R., & Bulun, S. E. (2001). Estrogen production and action. *Journal of the American Academy of Dermatology, 45*(3 Suppl), S116–S124.
105. Scherbakov, A. M., Krasilnikov, M. A., & Kushlinskii, N. E. (2013). Molecular mechanisms of hormone resistance of breast cancer. *Bulletin of Experimental Biology and Medicine, 155*, 384–395.
106. Williams, C., & Lin, C. Y. (2013). Estrogen receptors in breast cancer: Basic mechanism and clinical implications. *E Cancer Medical Science, 7*, 370.
107. Ramaswamy, S., et al. (2001). Multiclass cancer diagnosis using tumor gene expression signatures. *Proceedings of the National Academy of Sciences, 98*, 15149–15154.
108. Björnström, L., & Sjöberg, M. (2005). Mechanisms of estrogen receptor signaling: Convergence of genomic and nongenomic actions on target genes. *Molecular Endocrinology, 19*, 833–842.
109. Marchia, C., & Reis-Filho, J. S. (2008). Molecular diagnosis in breast cancer. *Diagnostic Histopathology, 14*, 202–213.

110. Barnes, D. M., & Hanby, A. M. (2001). Estrogen and progesterone receptors in breast cancer, past, present and future. *Histopathology, 38*, 271–274.
111. Hawkins, R. A., Roberts, M. M., & Forrest, A. P. M. (1980). Estrogen receptors and breast cancer. Current status. *The British Journal of Surgery, 67*, 162–165.
112. Harvey, J. M., Clark, G. M., Osborne, C. K., & Allred, D. C. (1999). Estrogen receptor status by immunohistochemistry is superior to the ligand-binding assay for predicting response to adjuvant endocrine therapy in breast cancer. *Journal of Clinical Oncology, 17*, 1474–1481.
113. Regitnig, P., Reiner, A., Dinges, H. P., Hofler, G., Muller-Holzner, E., Lax, S., Obrist, P., Rudas, M., & Quehenberger, F. (2002). Quality assurance for detection of estrogen and progesterone receptor by immunohistochemistry in Austrian pathology laboratories. *Virchows Archiv, 441*, 328–334.
114. Campbell, E. J., Tesson, M., Doogan, F., Mohammed, Z. M. A., Mallon, E., & Edwards, J. (2016). The combined endocrine receptor in breast cancer, a novel approach to traditional hormone receptor interpretation and a better discriminator of outcome than ER and PR alone. *British Journal of Cancer, 115*, 967–973.
115. Akiyama, T., Sudo, C., et al. (1986). The product of the human C-Erbb-2 gene – a 185-kilodalton glycoprotein with tyrosine kinase-activity. *Science, 232*, 1644–1646.
116. Shih, C., Padhy, L. C., et al. (1981). Transforming genes of carcinomas and neuroblastomas introduced into mouse fibroblasts. *Nature, 290*, 261–264.
117. Schechter, A. L., Stern, D. F., et al. (1984). The *neu* oncogene: An *erb-B*-related gene encoding a 185,000-Mr tumour antigen. *Nature, 312*, 513–516.
118. Muleris, M., Almeida, A., et al. (1997). Assignment of v-erb-b2 avian erythroblastic leukemia viral oncogene homolog 2 (ERBB2) to human chromosome band 17q21.1 by in situ hybridization. *Cytogenetics and Cell Genetics, 76*, 34–35.
119. King, C. R., Kraus, M. H., et al. (1985). Amplification of a novel v-erbB-related gene in a human mammary carcinoma. *Science, 229*, 974–976.
120. Kraus, M. H., Issing, W., et al. (1989). Isolation and characterization of ERBB3, a third member of the ERBB/epidermal growth factor receptor family: Evidence for overexpression in a subset of human mammary tumors. *Proceedings of the National Academy of Sciences of the United States of America, 86*, 9193–9197.
121. Plowman, G. D., Culouscou, J. M., et al. (1993). Ligand-specific activation of HER4/p180erbB4, a fourth member of the epidermal growth factor receptor family. *Proceedings of the National Academy of Sciences of the United States of America, 90*, 1746–1750.
122. Robinson, D. R., Wu, Y. M., et al. (2000). The protein tyrosine kinase family of the human genome. *Oncogene, 19*, 5548–5557.
123. Herbst, R. S. (2004). Review of epidermal growth factor receptor biology. *International Journal of Radiation Oncology, Biology, Physics, 59*(2 Suppl), 21–26.
124. Hubbard, S. R., & Till, J. H. (2000). Protein tyrosine kinase structure and function. *Annual Review of Biochemistry, 69*, 373–398.
125. Sundaresan, S., Roberts, P. E., et al. (1998). Biological response to ErbB ligands in nontransformed cell lines correlates with a specific pattern of receptor expression. *Endocrinology, 139*, 4756–4764.
126. Sliwkowski, M. X. (2003). Ready to partner. *Nature Structural Biology, 10*, 158–159.
127. Garrett, T. P. J., McKern, N. M., et al. (2002). Crystal structure of a truncated epidermal growth factor receptor extracellular domain bound to transforming growth factor α. *Cell, 110*, 763–773.
128. Jones, R. B., Gordus, A., et al. (2006). A quantitative protein interaction network for the ErbB receptors using protein microarrays. *Nature, 439*, 168–174.
129. Roy, V., & Perez, E. A. (2009). Beyond trastuzumab: Small molecule tyrosine kinase inhibitors in HER-2-positive breast cancer. *The Oncologist, 14*, 1061–1069.
130. Olayioye, M. A., Neve, R. M., et al. (2000). The ErbB signaling network: Receptor heterodimerization in development and cancer. *The EMBO Journal, 19*, 3159–3167.

131. Pinkas-Kramarski, R., Soussan, L., Waterman, H., Levkowitz, G., Alroy, I., Klapper, L., Lavi, S., Seger, R., Ratzkin, B. J., Sela, M., et al. (2006). Diversification of Neu differentiation factor and epidermal growth factor signaling by combinatorial receptor interactions. *The EMBO Journal, 15*, 2452–2467.
132. Zhang, W., & Liu, H. T. (2002). MAPK signal pathways in the regulation of cell proliferation in mammalian cells. *Cell Research, 12*, 9–18.
133. Pearson, G., Robinson, F., et al. (2001). Mitogen-activated protein (MAP) kinase pathways: Regulation and physiological functions. *Endocrine Reviews, 22*, 153–183.
134. West, K. A., Castillo, S. S., et al. (2002). Activation of the PI3K/Akt pathway and chemotherapeutic resistance. *Drug Resistance Updates, 5*, 234–248.
135. Yano, T., Ohtsu, A., Boku, N., Hashizume, K., Nakanishi, M., & Ochiai, A. (2006). Comparison of HER2 gene amplification assessed by fluorescence in situ hybridization and HER2 protein expression assessed by immunohistochemistry in gastric cancer. *Oncology Reports, 15*, 65–71.
136. Neve, R. M., Lane, H. A., et al. (2001). The role of overexpressed HER2 in transformation. *Annals of Oncology, 12*, 9–13.
137. Howe, L. R., Chang, S. H., et al. (2005). HER2/neu-induced mammary tumorigenesis and angiogenesis are reduced in cyclooxygenase-2 knockout mice. *Cancer Research, 65*, 10113–10119.
138. Mitri, Z., Constantine, T., et al. (2012). The HER2 receptor in breast cancer: Pathophysiology, clinical use, and new advances in therapy. *Chemotherapy Research and Practice, 2012*, 743193.
139. Maxwell, W. L. (1994). *Wheater's functional histology–a text and color atlas by H.G. Burkitt, B. Young, and J.W. Heath* (3rd ed., 416 pp). Edinburgh: Churchill Livingston.
140. Allred, D. C., et al. (1992). Overexpression of HER-2/neu and its relationship with other prognostic factors change during the progression of in situ to invasive breast cancer. *Human Pathology, 23*, 974–979.
141. Collins, L. C., & Schnitt, S. J. (2005). HER2 protein overexpression in estrogen receptor-positive ductal carcinoma in situ of the breast: Frequency and implications for tamoxifen therapy. *Modern Pathology, 18*, 615–620.
142. Hoff, E. R., Tubbs, R. R., et al. (2002). HER2/neu amplification in breast cancer: Stratification by tumor type and grade. *American Journal of Clinical Pathology, 117*, 916–921.
143. Ramos-Vara, J. A. (2005). Technical aspects of immunohistochemistry. *Veterinary Pathology, 42*, 405–426.
144. Tan, M., & Yu, D. (2007). Molecular mechanisms of ErbB2-mediated breast cancer chemoresistance. *Advances in Experimental Medicine and Biology, 608*, 119–129.
145. Creighton, C. J. (2012). The molecular profile of luminal B breast cancer. *Biologics, 6*, 289–297.
146. Fendly, B. M., Winget, M., et al. (1990). Characterization of murine monoclonal-antibodies reactive to either the human epidermal growth-factor receptor or Her2 Neu gene-product. *Cancer Research, 50*, 1550–1558.
147. Goldenberg, M. M. (1999). Trastuzumab, a recombinant DNA-derived humanized monoclonal antibody, a novel agent for the treatment of metastatic breast cancer. *Clinical Therapeutics, 21*, 309–318.
148. Scheuer, W., Friess, T., et al. (2009). Strongly enhanced antitumor activity of trastuzumab and pertuzumab combination treatment on HER2-positive human xenograft tumor models. *Cancer Research, 69*, 9330–9336.
149. Zeglinski, M., Ludke, A., et al. (2011). Trastuzumab-induced cardiac dysfunction: A 'dual-hit'. *Experimental and Clinical Cardiology, 16*, 70–74.
150. Pritchard, K. I., Shepherd, L. E., et al. (2006). HER2 and responsiveness of breast cancer to adjuvant chemotherapy. *The New England Journal of Medicine, 354*, 2103–2111.
151. Hung, M. C., & Lau, Y. K. (2001). Basic science of HER-2/neu and breast cancer. *Diagnostic Molecular Pathology, 10*, 139–152.

152. Suo, Z., Risberg, B., Karlsson, M. G., Villman, K., Skovlund, E., & Nesland, J. M. (2002). The expression of EGFR family ligands in breast carcinomas. *International Journal of Surgical Pathology, 10*, 91–99.
153. Zhao, J., Wu, R., Au, A., Marquez, A., Yu, Y., & Shi, Z. (2002). Determination of HER2 gene amplifi cation by chromogenic in situ hybridization (CISH) in archival breast carcinoma. *Modern Pathology, 15*, 657–665.
154. Rhodes, A., Jasani, B., Anderson, E., Dodson, A. R., & Balaton, A. J. (2002). Evaluation of HER2/neu immunohistochemical assay sensitivity and scoring in formalin-fixed and paraffin-processed cell lines and breast tumors. A comparative study involving results from laboratories in 21 countries. *American Journal of Clinical Pathology, 118*, 408–417.
155. Smith, K. L., Robbins, P. D., Dawkins, H. J., Papadimitriou, J. M., Redmond, S. L., Carrello, S., Harvey, J. M., & Sterrett, G. F. (1994). C-erbB-2 amplification in breast cancer. Detection in formalin-fixed paraffin-embedded tissue by in situ hybridization. *Human Pathology, 25*, 413–418.
156. Yerushalmi, R., et al. (2010). Ki67 in breast cancer: Prognostic and predictive potential. *The Lancet Oncology, 11*, 174–183.
157. Scholzen, T., & Gerdes, J. (2000). The Ki-67 protein: From the known and the unknown. *Journal of Cellular Physiology, 182*, 311–322.
158. Rao, C., Shetty, J., & Prasad, K. H. (2013). Immunohistochemical profile and morphology in triple -negative breast cancers. *Journal of Clinical and Diagnostic Research, 7*, 1361–1365.
159. Ribelles, N., et al. (2013). Pattern of recurrence of early breast cancer is different according to intrinsic subtype and proliferation index. *Breast Cancer Research, 15*, R98.
160. Sørlie, T., Perou, C. M., et al. (2006). Gene expression profiles do not consistently predict the clinical treatment response in locally advanced breast cancer. *Molecular Cancer Therapeutics, 5*, 2914–2918.
161. Gurvits, N., et al. (2017). Separase is a marker for prognosis and mitotic activity in breast cancer. *British Journal of Cancer, 117*, 1383–1391.
162. Tang, G., et al. (2011). Comparison of the prognostic and predictive utilities of the 21-gene Recurrence Score assay and Adjuvant! For women with node-negative, ER-positive breast cancer: Results from NSABP B-14 and NSABP B-20. *Breast Cancer Research and Treatment, 127*, 133–142.
163. van't Veer, L. J., et al. (2002). Gene expression profiling predicts clinical outcome of breast cancer. *Nature, 415*, 530–536.
164. van de Vijver, M. J., et al. (2002). A gene expression signature as a predictor of survival in breast cancer. *The New England Journal of Medicine, 347*, 1999–2009.
165. Dowsett, M., et al. (2013). Comparison of PAM50 risk of recurrence score with oncotype DX and IHC4 for predicting risk of distant recurrence after endocrine therapy. *Journal of Clinical Oncology, 31*, 2783–2790.
166. Pleasance, E. D., et al. (2010). A comprehensive catalogue of somatic mutations from a human cancer genome. *Nature, 463*, 191–196.
167. Pleasance, E. D., et al. (2010). A small-cell lung cancer genome with complex signatures of tobacco exposure. *Nature, 463*, 184–190.
168. Banerji, S., et al. (2012). Sequence analysis of mutations and translocations across breast cancer subtypes. *Nature, 486*, 405–409.
169. Ellis, M. J., et al. (2012). Whole-genome analysis informs breast cancer response to aromatase inhibition. *Nature, 486*, 353–360.
170. Shah, S. P., et al. (2012). The clonal and mutational evolution spectrum of primary triple-negative breast cancers. *Nature, 486*, 395–399.
171. Nik-Zainal, S., et al. (2016). Landscape of somatic mutations in 560 breast cancer whole-genome sequences Hutchinson L. Genetics: Defining driver mutations in the genomic landscape of breast cancer. *Nature Reviews. Clinical Oncology, 13*, 327.
172. Wu, Y. M., et al. (2013). Identification of targetable FGFR gene fusions in diverse cancers. *Cancer Discovery, 3*, 636–647.

173. Giacomini, C. P., et al. (2013). Breakpoint analysis of transcriptional and genomic profiles uncovers novel gene fusions spanning multiple human cancer types. *PLoS Genetics, 9*, e1003464.
174. Robinson, D. R., et al. (2011). Functionally recurrent rearrangements of the MAST kinase and Notch gene families in breast cancer. *Nature Medicine, 17*, 1646–1651.
175. Khurana, E., et al. (2013). Integrative annotation of variants from 1092 humans: Application to cancer genomics. *Science, 342*, 1235587.
176. West, J. A., et al. (2014). The long noncoding RNAs NEAT1 and MALAT1 bind active chromatin sites. *Molecular Cell, 55*, 791–802.
177. Nik-Zainal, S., et al. (2012). Mutational processes molding the genomes of 21 breast cancers. *Cell, 149*, 979–993.
178. Alexandrov, L. B., et al. (2013). Signatures of mutational processes in human cancer. *Nature, 500*, 415–421.
179. Stratton, M. R., Campbell, P. J., & Futreal, P. A. (2009). The cancer genome. *Nature, 458*, 719–724.
180. Siravegna, G., et al. (2017). Integrating liquid biopsies into the management of cancer. *Nature Reviews. Clinical Oncology, 14*, 531–548.
181. Pink, R. C., et al. (2017). Royal Society Scientific Meeting: extracellular vesicles in the tumor microenvironment. *Philosophical Transactions of the Royal Society of London. Series B, Biological Sciences, 373*, 2017.
182. Kalluri, R. (2016). The biology and function of exosomes in cancer. *The Journal of Clinical Investigation, 126*, 1208–1215.
183. Théry, C. (2011). Exosomes: secreted vesicles and intercellular communications. *F1000 Biology Reports, 3*, 15.
184. Menck, K., et al. (2015). Tumor-derived microvesicles mediate human breast cancer invasion through differentially glycosylated EMMPRIN. *Journal of Molecular Cell Biology, 7*, 143–153.
185. Menck, K., et al. (2017). Characterization of tumor-derived microvesicles in cancer patients' blood and correlation with clinical outcome. *Journal of Extracellular Vesicles, 6*, 1340745.
186. Minciacchi, V. R., et al. (2015). Large oncosomes contain distinct protein cargo and represent a separate functional class of tumor-derived extracellular vesicles. *Oncotarget, 6*, 11327–11341.
187. Minciacchi, V. R., et al. (2017). MYC mediates large oncosome-induced fibroblast reprogramming in prostate cancer. *Cancer Research, 77*, 2306–2317.
188. Zweemer, A. J., et al. (2017). Apoptotic bodies elicit Gas6-mediated migration of AXL-expressing tumor cells. *Molecular Cancer Research, 15*, 1656–1666.
189. Akyüz, N., et al. (2017). T-cell diversification reflects antigen selection in the blood of patients on immune checkpoint inhibition and may be exploited as liquid biopsy biomarker. *International Journal of Cancer, 140*, 2535–2544.
190. Beausang, J. F., et al. (2017). T cell receptor sequencing of early-stage breast cancer tumors identifies altered clonal structure of the T cell repertoire. *Proceedings of the National Academy of Sciences of the United States of America, 114*, E10409–E10417.
191. Adams, D. L., et al. (2014). Circulating giant macrophages as a potential biomarker of solid tumors. *Proceedings of the National Academy of Sciences of the United States of America, 111*, 3514–3519.
192. Cima, I., et al. (2016). Tumor-derived circulating endothelial cell clusters in colorectal cancer. *Science Translational Medicine, 8*, 345.
193. Lin, P. P., et al. (2017). Comprehensive in situ co-detection of aneuploid circulating endothelial and tumor cells. *Scientific Reports, 7*, 9789.
194. Sol, N., & Wurdinger, T. (2017). Platelet RNA signatures for the detection of cancer. *Cancer Metastasis Reviews, 36*, 263–272.
195. Best, M. G., et al. (2017). Swarm intelligence-enhanced detection of non–small-cell lung cancer using tumor-educated platelets. *Cancer Cell, 32*, 238–252.

196. Huang, G., et al. (2017). Integrating multiple fitting regression and Bayes decision for cancer diagnosis with transcriptomic data from tumor-educated blood platelets. *The Analyst, 142*, 3588–3597.
197. The Circulating Cell-free Genome Atlas Study (CCGA). Available at: https://clinicaltrials.gov/ct2/show/NCT02889978. Accessed 7 Sept 2016.
198. The STRIVE Study: Breast Cancer Screening Cohort. Available at: https://clinicaltrials.gov/ct2/show/NCT03085888. Accessed 21 Mar 2017.

Chapter 2
Breast Cancer Recurrence and Survival

Abstract Recurrence of breast cancer is a major cause of the disease-specific death. In general, it occurs within 5 years or even up to 10–20 years after surgery. It is no doubt that surgery follow-up by adjuvant endocrine therapy and chemotherapy has shown to be effective in reducing both local and distant metastasis of breast cancer. An important goal in medical oncology is to identify those women at high risk of recurrence. Prediction of recurrence has been the quest of the pathologist by using morphological phenotypes predictor of early disease recurrence. A better understanding of the cellular and molecular mechanism of invasion and metastasis is pointing cancer stem cells as the main player for shortening patient survival. Therefore, detection of circulating stem cells is one of the possible ways to detect earlier and to have a tool for better follow-up of the behavior of the cancer cells. The technological advances in single cell analysis and the identification of circulating tumor cells (CTCs) under the general concept of liquid biopsy is probably the methodology that will change the way that pathology is practiced today and the oncologist will use for determining not only the adequate treatment but also how the cancer patients respond to treatment and providing the adequate queues when drug resistance and relapse is emerging.

Keywords Recurrence of breast cancer · Adjuvant endocrine therapy and chemotherapy · Cancer stem cells

2.1 Cancer Recurrence

Recurrence of breast cancer is a major cause of the disease-specific death. In general, it occurs within 5 years or even up to 10–20 years after surgery [1–4]. It is no doubt that surgery follow-up by adjuvant endocrine therapy and chemotherapy has shown to be effective in reducing both local and distant metastasis of breast cancer [5]. Women with hormone receptor-positive breast cancer are the best responder, but relapses still occur 5 years after diagnosis [5]; however in hormone receptor-negative disease, the risk of recurrence is confined mostly to the first 5 years after diagnosis, and relapse rates fall rapidly thereafter [6, 7]. In hormone receptor-positive breast cancer, the extension of the hormone therapy for more than 5 years

J. Russo, *The Future of Prevention and Treatment of Breast Cancer*,
https://doi.org/10.1007/978-3-030-72815-1_2

either with Tamoxifen or aromatase inhibitors has prolonged the free interval period of the disease [8–10].

An important goal in medical oncology is to identify those women at high risk of recurrence. Prediction of recurrence has been the quest of the pathologist by using morphological phenotypes predictor of early disease recurrence (see Chap. 1). A better understanding of the cellular and molecular mechanism of invasion and metastasis is pointing cancer stem cells as the main player for shortening patient survival [11, 12]. Therefore detection of circulating stem cells is one of the possible ways to detect earlier and to have a tool for better follow-up of the behavior of the cancer cells [13, 14, and Sect. 2.2.]

2.2 Markers of Cancer Recurrence

In the previous chapter we have described the main prognostic markers used in the prediction of recurrence or free disease interval in breast cancer. It is important to remind here that tumor size, stage, and nodal involvement are still relevant for both estrogen receptors (ER)-positive and ER-negative breast cancer. The usefulness of tumor grade, ER, PgR, Ki-67, as well as other markers like ERB 2 has been described in Chap. 1. The use of molecular markers and the molecular tests are also widely used on the last 5 years (see Chap. 1).

The technological advances in single cell analysis and the identification of circulating tumor cells (CTCs) under the general concept of liquid biopsy [15–17] is probably the methodology that will change the way that pathology is practiced today and the oncologist will use for determining not only the adequate treatment but also how the cancer patients respond to treatment and providing the adequate queues when drug resistance and relapse is emerging.

CTCs are linked to metastatic relapse and can be used as a prognostic marker not only for breast but also for other cancers [18–23]. Using immunodeficient mice it has been possible to identify properties of the stem cells that are circulating in blood [24] and establish metastatic properties, for example, for the luminal cancer cells expressing the phenotypes of EPCAMlow, METhigh, CD47high, and CD44high [24]. An important observation has been that cells circulating in cluster have more metastatic potential than single circulating cells [16, 17]. However not all cells in cluster are pure cancer cells but other blood components such as platelets and leukocytes [25–27] that are facilitating the growth of metastatic loci at a distant site [16, 28]. This observation has led to the discovery of plakoglobin, an important component of desmosomes [16, 29], and cells with unregulated levels of plakoglobin show lower motility, while those with low plakoglobin levels display high metastatic potential [30–33]. Although association between plakoglobin and malignancies still remains controversial [34–36], in the study of Lu et al. [37] using plakoglobin expression and breast cancer in which patients were classified into three subgroups based on the tertile distribution of plakoglobin expression, low ($-3.570 \leq$ plakoglobin <-0.508), medium ($-0.508 \leq$ plakoglobin < 0.208), and high

($0.208 \leq$ plakoglobin ≤ 2.726), the Kaplan–Meier survival curves showed that patients with low plakoglobin expression had significantly better distant metastasis-free survival ($p = 0.02182$) and overall survival ($p < 0.00001$), and borderline significantly superior relapse-free survival ($p = 0.05008$) compared to those with high one. After adjusting the potential confounding factors (which include estrogen status (ER), lymph node status, tumor grade, patient age, and study using multivariate Cox proportional hazard regression analyses), again, patients with low plakoglobin expression had significantly reduced risks of distant metastasis ($p = 0.01$), death ($p = 1 \times 10{-}5$), and relapse ($p = 0.001$) compared to those with high one. Patients with low plakoglobin expression had approximately 75% reduced death risks, followed by 50% reduced relapse risks and 40% reduced distant metastatic risks, and that those with the medium one had approximately 50% reduced death risks ($p = 0.006$), followed by 30% reduced relapse risks ($p = 0.05$) and 20% reduced distant metastasis risks ($p = 0.4$). These findings suggest that plakoglobin expression is an independent prognostic factor in the patients with breast cancer; particularly for overall survival, those with low plakoglobin expression had superior survival than those with the high one [37].

CD44+CD24-/lowLineage was used for identifying the breast cancer cells as earlier as 2003 [38]. Later work has pointed the activity of aldehyde dehydrogenase 1 (ALDH1) as assessed by the Aldefluor assay resulted to be a specific indicator for identifying, isolating, and tracking human breast cancer stem cells (CSCs) [39].

The ALDH1A subfamily comprises three isoforms (ALDH1A1, ALDH1A2, and ALDH1A3), which synthesize retinoic acid (RA) from the retina and are crucial regulators for the RA signaling pathway. These enzymes have a high affinity for the oxidation of both all-trans- and 9-cis-retinal and thereby serve to regulate the self-renewal and differentiation of normal stem cells and CSCs [40]. The prognostic value of ALDH1A1 for breast cancer remains controversial despite numerous independent studies. Ginestier et al. demonstrated that ALDH1A1 expression detected by immunostaining correlated with poor patient prognosis [39]. The prognostic value of ALDH1A1 expression is age dependent and can be observed only in patients aged <65 years [40]. Using a retrospective collection of 321 node-negative and 318 node-positive breast cancer patients with a mean follow-up time of 12.6 years, Neumeister et al. found that ALDH1A1 expression alone does not significantly predict therapeutic outcome [41]. Liu et al. [42] performed a systematic review and a meta-analysis to assess the robustness of the relationship between ALDH1A1 expression and clinic-pathologic parameters/outcomes in breast cancer patients. This author used a total of 16 studies [39, 41, 43–55], and the total number of patients was 3274, including 921 cases ALDH1A1+ breast cancer and a 2353 controls. In this meta-analysis it showed a significant association between ALDH1A1 expression and tumor size, histological grade, LNM, and the expression of ER, PR, and HER2. Specifically, higher ALDH1A1 expression means greater tumor size, higher SBR grade, greater possibility of LNM, higher expression of HER2, and lower expression of ER and PR. Subsequently, they performed a subcategory analysis according to different cutoff values of ALDH1A1 expression (>5% and >0%/1%

subgroups), different regions the patients originated from (America-Europe, Asia, and Africa subgroups), different clinical stages of the patients [No assessment (NA) and ≥ stage II subgroups], different antibodies used in the studies (BD subgroup and Abcam subgroup), and types of surgery for patients [Surgery, Part surgery, and No screened (NS) subgroups] [42]. In the subgroup analysis based on the cutoff value, they found that ALDH1A1 expression is positively correlated with histological grade and negatively correlated with the expression of ER/PR, which is consistent with the results derived from overall analysis. At the same time, greater tumor size and higher expression of HER2 in the ALDH1A1 positive group could be found in the subgroup studies with cutoff values >0% or 1%. However, LNM status is not correlated with ALDH1A1 expression regardless of cutoff value. This meta-analysis indicates that ALDH1A1 is an important predictor of the progression and poor survival of breast cancer patients. The results suggest that the analysis of ALDH1A1 expression in breast cancer not only provides a better understanding of the relationship between breast tumorigenesis and cancer genomics but may also be beneficial for the design of treatment and the assessment of the prognosis of patients [42].

CXCR4 can signify invasion and metastasis in several cancers, including breast cancer [56]. The chemokine receptor CXCR4 is a 352-amino acid rhodopsin-like G protein-coupled receptor (GPCR) that selectively binds the CXC chemokine stromal cell-derived factor 1 (SDF-1), also known as CXCL12. This chemokine receptor has been identified to play a crucial role in a number of biological processes, such as trafficking and homeostasis of immune cells such as T lymphocytes [57], and the CXCL12/CXCR4 axis is known to be important in the enhancement of stem cell homing in tissue regeneration [58]. Importantly it has also been identified in breast cancer [56]. It is believed that its involvement in tumor dissemination and metastasis is through promoting its trans-endothelial migration at the primary site [59]. Further evidence indicates that CXCR4 not only affects breast cancer metastasis but also promotes the survival and proliferation of breast cancer cells through increasing the number of blood vessels in tumors [60]. In a meta-analysis performed by Zhang et al. [61], they determined the relationship between CXCR4 and several clinic-pathological features of breast cancer. A total of 13 studies were included in the meta-analysis consisting of approximately 3865 participants with a median of 297 (from 54 to 1699) per study. These studies evaluated the expression of CXCR4 and evaluation parameters for breast cancer, based on either clinic-pathological features or prognostic factors. Among them, 7 were from the USA, 1 was from Canada, 3 were from China, 1 was from Japan, and 1 was from Korea. The study showed a significant correlation between CXCR4 expression and the 5-year OS and DFS rates in breast cancer patients. CXCR4 expression was also associated with lymph node status and distant metastasis. Thus, CXCR4 could have prognostic significance for patients with breast cancer. Recent studies indicated that small antagonists, such as AMD3100, inhibited the primary tumor and metastasis in animal models of breast cancer [62]. Hence, CXCR4 antagonists could have a significant therapeutic impact on primary and metastatic breast cancer by disrupting tumor vasculature in the microenvironment [63]. Furthermore, inhibiting CXCR4 in tumor cells has the potential to induce growth arrest or apoptosis and to prevent invasion and metastasis [63]. Hence, in addition to being a prognostic marker, CXCR4 could also be an anti-cancer therapy target. Recent preclinical studies in mouse models of leukemia have provided proof of concept for the greater benefits of combining CXCR4 inhibition with conventional chemotherapy relative to chemotherapy treatment alone [64, 65]. We infer that in the case of breast cancer, a CXCR4 inhibitor could improve survival and prognosis.

2.3 Cancer Survival in Sporadic Breast Cancer

Despite local control by surgery and radiotherapy, many women who present with apparently localized breast cancer will ultimately relapse and die from distant metastases [66]. This has been the rationale to advise the use of adjuvant chemotherapy to reduce the risk of distant metastases and improve overall survival [67–69]. In a systematic review and meta-analysis to determine whether delay in initiation of adjuvant chemotherapy decreases its effectiveness [70], a statistically significant association was found between delay in a 4-month range in adjuvant chemotherapy and survival, indicating that treatment must start as soon as possible after surgery, when the burden of residual tumor cells is at its lowest [71]. This means that the eradication of a cancer by chemotherapy is inversely related to the burden of clonogenic tumor cells and that delay in initiating adjuvant treatment provides the opportunity for tumor cell proliferation [71] and accelerates the growth of micro-metastases [72, 73]. This observation on the effect of delay and reduction in survival was already demonstrated [74], and a 4-week delay in AC is associated with increase in the relative risk of death of approximately 5%. It is true as speculated by the authors of this meta-analysis [70] that a short delay in AC has a relatively small impact on the prognosis of the individual patient, the potential societal impact of delays in AC is extremely important, and across the USA, this would save 1500 lives. The diagnosis of invasive breast cancer with mammographic screening has been reported in a survival gain at 5 years from the diagnosis [75].

Adjuvant chemotherapy (CT) reduces the relative risk (RR) of death in early breast cancer (BC) by about 20% when using anthracycline-based schedules, and by about a further 15% with the addition of taxanes [75]. The absolute gain in overall survival (OS) is slightly higher in estrogen-receptor-negative (ERs) BC. Adjuvant dose-dense (DD) CT aims to achieve maximum tumor death by delivering the drugs over a shorter period of time, but not by increasing the dosage (e.g., every 2 weeks instead of the classic every 3-weeks schedule). Theoretically, the growth of tumor cells follows a mathematically predictable growth curve, with rapid initial growth (when the cells are more sensitive to cytotoxics) followed by a plateau, particularly in the post-operative setting [76]. This theory is the rationale for the use of DD-CT, with the goal being to achieve a higher rate of cell destruction. Dose-dense (DD) chemotherapy (CT) aimed at achieving a higher rate of cancer cell destruction has been adopted as an adjuvant therapy in high-risk breast cancer (BC), with the goal being to improve outcomes. In a meta-analysis [77] in which a total of 8 phase III trials encompassing 17,188 randomized patients has shown that the patients who received DD-CT had better overall survival (OS: HR 0.86, 95% confidence interval [CI] 0.79–0.93, P = 0.0001) and disease-free survival (DFS: HR 0.84, 95% CI 0.77–0.91, P < 0.0001) than those on the conventional schedule. A statistically significant OS benefit was observed in patients with hormone receptor-negative (ER-) tumors (HR 0.8, P = 0.002), but not in those with ER-positive BC (HR 0.93, 95% CI 0.82–1.05; P = 0.25). DD-CT leads to better OS and DFS, particularly in women with ER-early BC. These results suggest that the DD strategy should be the standard care offered to high-risk ER-BC patients.

2.4 Cancer Survival in BRCA1 and BRCA2 Carriers in Breast Cancer

BRCA1 and BRCA2 are two distinct tumor suppressor genes controlling the DNA repair processes [78]. The lifetime risk of breast cancer among BRCA mutation carriers is 45–80% [79]. Several studies have investigated prognoses among BRCA mutation carriers and no carriers. Multiple studies have demonstrated that BRCA1-related breast cancers were more likely to be triple-negative, which was correlated with poor prognosis [80–82].

A meta-analysis published in 2010 also linked BRCA1 mutation to decreased overall (OS) and progression-free survival (PFS) among patients with breast cancer, although BRCA2 mutation was not associated with differential survival outcomes [83]. A multi-country study failed to show differences in breast cancer survival rates associated with BRCA mutations [84]. Cortesi and colleagues [85] reported favorable 10-year breast cancer outcomes among BRCA1 carriers. Meanwhile, a large-scale study involving 2967 patients demonstrated adverse effect of BRCA2 mutations on breast cancer-specific survival [86]. In a meta-analysis published in 2015 [87], 13 studies conducted in 12 countries (Poland, Italy, Israel, Norway, the UK, the Netherlands, Canada, France, Finland, the USA, Australia, and Germany) and published between 2000 and 2013 with a total of 10,016 patients (ranging from 85 to 3345 per study) were included, among them 890 BRCA1 mutation carriers, 342 BRCA2 mutation carriers, and 8784 non-carriers. Briefly they found from 11 studies a worse breast cancer OS among BRCA1 mutation carriers than non-carriers. Because a significant heterogeneity was found across the studies (I2 = 59%; P = 0.007), the pooled HR of 1.5 (95% CI, 1.11–2.04) was calculated on the basis of a random-effects model. The result of a meta-analysis of five studies comparing breast cancer OS between BRCA2 carriers and non-carriers indicated no association between mutation and OS, with a pooled HR of 0.97 (95% CI, 0.78–1.22), and no significant heterogeneity was found across the studies (I2 = 0%). Eight and two studies were included to assess the association between BRCA1 or BRCA2 mutations and breast cancer PFS, respectively. Meta-analyses results did not reveal any association between these two mutations and PFS, with a pooled HR of 1.35 (95%

CI, 0.95–1.90) for BRCA1 mutation carriers and 0.95 (95% CI, 0.69–1.30) for BRCA2 mutation carriers. A random-effects model was used for the BRCA1 mutation meta-analysis due to a significant heterogeneity among studies (I2 = 55%, $P = 0.03$). No significant heterogeneity among studies was found in the meta-analysis of BRCA2 mutation [87].

The mechanism driving the association between BRCA1 or BRCA2 mutation and survival is not clear. It may be rooted in different functions of the BRCA genes in the dsDNA repair pathway. Research on breast cancer also indicated that BRCA1 mutation carriers had more p53 mutations, resulting in worse prognosis [83]. The different breast cancer outcomes observed among BRCA1 and BRCA2 carriers could also be due to their distinct clinical characteristics. As mentioned previously, BRCA1-related breast cancers were BRCA1/2-associated breast cancers account for about 25–30% of familial breast cancers, and for about 3% of all breast cancers [88, 89]. BRCA1-associated breast cancers differ from tumors not associated with BRCA mutations with respect to pathological features, e.g., they are more often estrogen receptor negative and high grade and have a higher frequency of somatic abnormalities in prognostically important genes such as P53 [90, 91]. The biological background of BRCA1/2 [92] and different pathological aspects of BRCA1-associated tumors support the hypothesis that patients carrying a BRCA1 and/or BRCA2 mutation might have a worse breast cancer prognosis compared to non-carriers. An impressive number of studies have already been conducted to address the association between BRCA1 and/or BRCA2 mutation carriership and breast cancer survival and the results were inconsistent [92]. In a study performed by van den Broek et al. [93], they identified 66 relevant studies concluding a moderate evidence for a worse unadjusted recurrence-free survival for BRCA1 mutation carriers. For BRCA1 and BRCA2 there was a tendency toward a worse breast cancer-specific and overall survival; however, results were heterogeneous and the evidence was judged to be indecisive. Surprisingly, only eight studies considered adjuvant treatment as a confounder or effect modifier while only two studies took prophylactic surgery into account. Adjustment for tumor characteristics tended to shift the observed risk estimates toward a relatively more favorable survival. The authors concluded that in contrast to currently held beliefs of some oncologists, current evidence does not support worse breast cancer survival of BRCA1/2 mutation carriers in the adjuvant setting; differences if any are likely to be small.

In a meta-analysis reported by Zhong et al. [87], among patients with breast cancer, BRCA1 mutation carriers had worse OS (HR, 1.50; 95% CI, 1.11–2.04) than non-carriers but were not significantly different from non-carriers in PFS. BRCA2 mutation was not associated with breast cancer prognosis. Valachis et al. [94] investigated the oncological safety of breast-conserving therapy BCT in BRCA-mutation carriers and the risk for contralateral breast cancer (CBC) compared with non-carriers, potential risk factors for ipsilateral breast recurrence (IBR) or CBC and grades these factors based on the level of evidence. In this meta-analysis there was no significant difference in IBR between carriers and controls (RR 1.45, 95% CI 0.98–2.14). However, a significant higher risk for IBR in BRCA-mutation carriers was observed in studies with a median follow-up ≥7 years (RR 1.51, 95%

CI 1.15–1.98). CBCs were significantly greater in carriers versus controls (RR 3.56, 95% CI 2.50–5.08). Use of adjuvant chemotherapy and oophorectomy was associated with a significantly lower risk for IBR in BRCA-mutation carriers. Three factors were associated with a lower risk for CBC in BRCA-mutation carriers: oophorectomy, use of tamoxifen, and age at first breast cancer. Based on current evidence, the use of BCT in BRCA-mutation carriers can be considered a reasonable option since it does not seem to increase the risk for IBR.

2.5 Markers Predicting Survival

The development of signature biomarkers has gained considerable attention in the past decade. Although the most well-known examples of biomarker panels stem from gene expression studies, proteomic panels are becoming more relevant, with the advent of targeted mass spectrometry-based methodologies. At the same time, the development of multigene prognostic classifiers for early stage breast cancer patients has resulted in a wealth of publicly available gene expression data from thousands of breast cancer specimens. There is a great amount of literature pointing the use of different markers for predicting survival of breast cancer. Some of them are indicating where the future research is going and I am summarizing below.

Tang et al. [95] reported a total of 41 articles including 27 types of miRNAs were found regarding prognostic biomarkers for breast cancer survival, of which, micRNA-21 (miR-21) was the most-studied specific miRNA that appeared repeatedly among the selected classifiers. For the studies evaluating miR-21's association with clinical outcomes, the median HR in the studies was 2.32 (interquartile range [IQR] = 1.04–3.40), and the pooled HR suggested that high expression of miR-21 has a negative impact on overall survival (OS: HR = 1.46; 95% CI, 1.25–1.70; $p < 0.05$) and disease/recurrence-free survival in breast cancer (HR = 1.49; 95% CI, 1.17–1.90; $p < 0.01$). They also found that higher expression levels of miR-210 significantly predicted poorer outcome, with median HR in the reported studies of 4.07 (IQR = 1.54–4.43) and a pooled HR of 2.94 (95% CI, 2.08–4.17; $p < 0.05$). In summary the data indicate that miRNAs show promising associations with prognosis in breast cancer. Moreover, specific miRNAs such as miR-21 and miR-210 can predict poor survival rates in breast cancer patients. Pan et al. also explore the role of microRNA [96] mainly microRNA-21 (miR-21) as a prognostic biomarker of breast cancer. However, controversy still remains. Pan et al.'s studies were based from PubMed and EMBASE through multiple search strategies. Data were extracted from studies comparing survival in breast cancer patients having higher miR-21 expression with those having lower expression. A meta-analysis was performed to clarify prognostic role of miR-21 in patients with breast cancer. Subgroup analysis was also performed according to patients' ethnicity. A total of 6 eligible articles comprising 951 cases were selected for this meta-analysis. The combined hazard ratios (HRs) and 95% confidence intervals (95% CIs) for overall survival (OS) were 2.11 (1.09–4.08) and for disease-free survival (DFS) was 1.6 (1.30–1.96). Subgroup analysis indicated high miR-21 expression was significantly associated with worse OS in Asian patients (HR = 4.39; 95% CI, 2.47–7.80) but not in non-Asian patients (HR = 1.18; 95% CI, 0.81–1.70). Sensitivity analysis revealed results of this meta-analysis were robust. Odds ratios (ORs) showed that miR-21 expression was closely associated with estrogen receptor (ER), progesterone receptor (PR), lymph node metastasis, histological grade, and HER2/neu. The pooled ORs and 95% CIs were 0.53 (0.35–0.80), 0.49 (0.32–0.74), 2.32 (1.54–3.50), 2.44 (1.58–3.75), and 4.29 (2.34–7.85), respectively. The results indicated that elevated miR-21 expression could potentially predict poor survival in patients with breast cancer.

The use of integrated transcriptome and proteome-based platforms to identify genes and proteins related to patient survival has also been explored using mass-spectrometry-based assay for the simultaneous quantification of 20 proteins in breast cancer tissue extracts. The relative expression levels of 20 potential biomarkers in a cohort of 96 tissue samples from patients with early stage breast cancer identified two proteins, KPNA2 and CDK1, which showed potential to discriminate between estrogen receptor-positive patients of high and low risk of disease recurrence. The role of these proteins in breast cancer prognosis requires more studies [97].

In the publication of Abravanel et al. [98], they reported that Notch signaling is acutely upregulated in tumor cells following HER2/neu pathway inhibition, that Notch signaling remains activated in a subset of dormant residual tumor cells that persist following HER2/neu downregulation, and as a consequence activation of Notch signaling accelerates tumor recurrence, and that inhibition of Notch signaling by either genetic or pharmacological approaches impairs recurrence in mice [98]. Consistent with these findings, meta-analysis of microarray data from over 4000 breast cancer patients revealed that elevated Notch pathway activity is independently associated with an increased rate of recurrence. Together, these results implicate Notch signaling in tumor recurrence from dormant residual tumor cells and provide evidence that dormancy is a targetable stage of breast cancer progression. Yuan et al. [99] also explored the value of Notch1 expression in breast cancer, but the results have ever been disputed. By using 21 studies involving 3867 patients, this meta-analysis revealed that the expression of Notch1 was significantly higher in breast cancer than in normal tissues (OR = 7.21; 95% CI, 4.7–11.07) and that higher Notch1 expression was associated with transition from ductal carcinoma in situ (DCIS) to invasive cancer (OR = 3.75; 95% CI, 1.8–7.78). Higher Notch1 activity was observed in the basal subtype of breast cancer (OR = 2.53; 95% CI, 1.18–5.43). Moreover, patients with Notch1 overexpression exhibited significantly worse overall and recurrence-free survival. The meta-analysis suggests that Notch inhibitors may be useful in blocking the early progression of DCIS and that the outcomes of clinical trials for Notch1-targeting therapeutics could be improved by the molecular stratification of breast cancer patients [99].

Yan et al. [100] studied the role of c-Met in breast cancer, and a meta-analysis of the impact of this gene was performed by searching published data. Yan and coworkers [100] analyzed overall survival (OS) or relapse-free survival (RFS) according to c-Met expression. The principal outcome measures were hazard ratios (HRs) for RFS or OS according to c-Met expression. Combined HRs were calculated using fixed- or random- effects models according to the heterogeneity. Twenty-one studies involving 6010 patients were used, and the impact of c-Met on RFS and OS was investigated in 12 and 17 studies, respectively. The meta-analysis results showed that c-Met overexpression significantly predicted poor RFS and OS in unselected breast cancer. Subgroup analysis indicated that c-Met overexpression was correlated with poor RFS and OS in Western patients, but was not associated with RFS or OS in Asian patients. C-Met was associated with poor OS in lymph node-negative breast cancer and with poor RFS in hormone-receptor-positive and triple-negative breast cancer, but was not associated with prognosis in human epidermal growth factor receptor (HER)-2-positive breast cancer. Yan et al. [100] concluded c-Met overexpression is an adverse prognostic marker in breast cancer, except among Asian and HER-2-positive patients.

Lin et al. [101] analyzed the role of Interleukin-6 (IL-6) that has been shown to promote tumor survival, metastasis, and angiogenesis, in addition to possessing antitumor activities. In light of the conflicting data, Lin et al. [101] selected studies describing the use of IL-6 as a prognostic factor for breast cancer and found that overall survival (OS), disease-free survival (DFS), and clinic pathologic features were associated. They studied 13 articles containing 3224 breast cancer patients, and the results showed that IL-6 expression was not associated with lymph node metastasis, tumor size, or histologic grade. Moreover, there was no correlation between IL-6 expression and DFS. However, the combined hazard ratio (95% confidence interval) for OS was 2.15 (1.46, 3.17). Sensitivity analysis further demonstrated that, for OS, the results of this meta-analysis were stable. A subgroup analysis showed that the source used to detect IL-6 levels may have altered the pooled results for OS. The authors concluded that taken together, these results indicate that IL-6 expression is associated with poor prognosis for breast cancer, and the prognostic role is affected by the source used to detect IL-6 levels.

In another study Kim et al. [102] studied the prognostic value of ErbB3 in human breast cancer based on the role of this receptor in drug resistance. This investigator evaluated the relationship between ErbB3 expression and survival of breast cancer via meta-analysis using 32 potentially relevant studies. Outcomes presented in searched literatures can be classified as disease-free survival (DFS), overall survival (OS), and progress-free survival (PFS) values. Meta-analysis was performed for each group. Results showed no statistically significant difference in survival. The overall hazard ratio of PFS, DFS, and OS of ErbB3 expression was 1.40 [95% confidence interval/CI (0.51, 3.83)], 1.07 [95% CI (0.82, 1.40)], and 1.15 [95% CI (0.91, 1.44)], respectively. Subgroup analysis according to ErbB2 receptor status, ErbB3 assessment methods (immunohistochemistry/IHC vs non-IHC), and analysis type (multivariate and univariate analysis) were performed. No significant association was found. Using various assessment methods and patient populations, our results revealed that there was no significant correlation between ErbB3 expression and breast cancer survival. Further studies on heterodimers of ErbB3 and other molecular markers involved in ErbB3 related pathway are merited.

Li et al. [103] studied survivin, an inhibitor of apoptosis protein as a potentially prognostic factor and therapeutic target in breast carcinoma, and the relevant articles were screened in PubMed and EMBASE databases. Patients' clinical characteristics, overall survival (OS), disease/recurrence-free survival (DFS/RFS), and positive expressed survivin rates were extracted for further analysis. Statistics extracted from Kaplan–Meier survival curves were calculated indirectly with methods developed by Parmar, Williamson, and Tierney. Multivariate Cox hazard regression analysis data were used directly in Stata 11.0. Pooled hazard ratio (HR) and 95% confidence interval (CI) were calculated to evaluate the prognostic role of survivin in breast cancer. Online literature search identified 23 articles containing 3259 breast cancer patients. The meta-analysis of all included studies about survival outcomes showed positive correlation between poor prognosis and survivin expression. Pooled HRs (95% CIs) for OS and DFS/RFS were 1.37 (1.12–1.68) and 1.34 (1.02–1.76), respectively. Subgroup analyses considering methods used to detect survivin (immunohistochemistry or not) and localization of survivin (whole, nuclear, or cytoplasm of the cell) were also conducted, and all the above analyses supported the stability of the prognostic role of survivin. In addition, the study revealed a significant association between survivin expression and lymph node metastasis (OR, 2.74; 95% CI, 1.27–5.93) or stage of breast cancer (OR, 2.01; 95% CI, 1.29–3.13). Positive expression of survivin demonstrated a significantly higher risk of recurrence and decreased OS rates in breast cancer.

2.6 Other Factors Influencing Recurrence and Survival

Many other factors are also influencing breast cancer survival; among them are physical activity, smoking, and diet. Some of the best examples are discussed below.

A report of Lahart et al. [104] showed the effect of lack of physical activity on the risk of developing breast cancer. In this meta-analysis was studied the association between physical activity and breast cancer recurrence and death. Twenty-two prospective cohort studies were eligible, and the average follow-up periods range from 4.3 to 12.7 years with 123,574 participants, 6898 all-cause deaths, and 5462 breast cancer outcomes (i.e., breast cancer-related deaths or recurrences). The average Newcastle-Ottawa score was six stars (range 4–8). Compared to those who reported low/no lifetime recreational pre-diagnosis physical activity, participants who reported high lifetime recreational pre-diagnosis physical activity levels had a significantly lower risk of all-cause (HR = 0.82; 95% CI, 0.70–0.96, $p < 0.05$) and breast cancer-related death (HR = 0.73; 95% CI, 0.54–0.98, $p < 0.05$). Significant risk reductions for all-cause and breast cancer-related death were also demonstrated for more recent pre-diagnosis recreational physical activity (HR = 0.73; 95% CI, 0.65–0.82, $p < 0.001$; and HR = 0.84; 95% CI, 0.73–0.97, $p < 0.05$, respectively), post-diagnosis physical activity (HR = 0.52; 95% CI, 0.43–0.64, $p < 0.01$; and HR = 0.59; 95% CI, 0.45–0.78, $p < 0.05$, respectively), and meeting recommended physical activity guidelines (i.e., $\geq$ 8 MET-h/wk) post-diagnosis (HR = 0.54; 95% CI, 0.38–0.76, $p < 0.01$; and HR = 0.67; 95% CI, 0.50–0.90, $p < 0.01$, respectively). However, there was evidence of heterogeneity across lifetime recreational pre- and post-diagnosis physical activity analyses. Both pre-diagnosis (lifetime and more recent combined) and post-diagnosis physical activity were also associated with reduced risk of breast cancer events (breast cancer progression, new primaries, and recurrence combined) (HR = 0.72; 95% CI, 0.56–0.91, $p < 0.01$; and HR = 0.79; 95% CI, 0.63–0.98, $p < 0.05$, respectively). The final conclusion of this study was an inverse relationship between physical activity and all-cause, breast cancer-related death and breast cancer events. The current meta-analysis supports the notion that appropriate physical activity may be an important intervention for reducing death and breast cancer events among breast cancer survivors.

The relation of smoking at diagnosis to breast cancer-specific and overall survival among 5892 women with invasive breast cancer treated in one Canadian center (1987–2008) [105] has shown that women were classified as never, former, or current smokers. Current smokers were further classified according to total, intensity, and duration of smoking. Deaths were identified through linkage to population mortality data. Cox proportional-hazards multivariate models were used. Compared with never smokers, current smokers at diagnosis had a slightly, but not statistically significant, higher breast cancer-specific mortality (hazard ratio = 1.15; 95% confidence interval (CI), 0.97–1.37). Among current smokers, breast cancer-specific mortality increased with total exposure to, intensity, and duration of smoking (all Ptrend <0.05). Compared to never smokers, breast cancer-specific mortality was 32–56% higher among heavy smokers (more than 30 pack years of smoking, more than 20 cigarettes per day or more than 30 years of smoking). Smoking at diagnosis was associated with an increased all-cause mortality rate. A meta-analysis of all studies showed a statistically significant, 33% increased mortality from breast cancer in women with breast cancer who are smokers at diagnosis compared to never smokers (hazard ratio = 1.33; 95% CI, 1.12–1.58). This study indicated that smoking at diagnosis is associated with a reduction of both overall and breast cancer-specific survival.

Vitamin D regulates expression of genes important in development and progression of breast cancer. The association of vitamin D with breast cancer outcomes among breast cancer patients is controversial. A systematic review and meta-analysis [106] were performed using an inverse-variance weighted fixed-effects model with Stata Version 12. Eight studies including 5691 patients were identified. Vitamin D deficiency was variably categorized across studies; a median of 36.8% of patients were classified as deficient. Low vitamin D levels were associated with a pooled hazard ratio of 2.13 (95% CI 1.64–2.78) and 1.76 (95% CIs 1.35–2.30) for recurrence (six studies) and death (four studies), respectively, with no evidence of significant heterogeneity across studies. There was potential evidence of a publication bias in studies examining associations with death (but not in those examining associations with recurrence). These findings support an association of low levels of vitamin D with increased risk of recurrence and death in early stage breast cancer patients. Another meta-analysis [107] examined the association between vitamin D status (based on diet and blood 25-hydroxyvitamin D (25(OH)D)) and breast cancer risk or mortality. A total of 30 prospective studies (nested case-control or cohort) were included for breast cancer incidence ($n = 24$ studies; 31,867 cases) or mortality

($n = 6$ studies; 870 deaths) among 6092 breast cancer patients. The pooled RRs of breast cancer incidence for the highest vs the lowest vitamin D intake and blood 25(OH)D levels were 0.95 (95% CI, 0.88–1.01) and 0.92 (95% CI, 0.83–1.02), respectively. Among breast cancer patients, high blood 25(OH)D levels were significantly associated with lower breast cancer mortality (pooled RR = 0.58; 95% CI, 0.40–0.85) and overall mortality (pooled RR = 0.61; 95% CI, 0.48–0.79). There was no evidence of heterogeneity and publication bias suggesting that high vitamin D status is weakly associated with low breast cancer risk but strongly associated with better breast cancer survival [107, 108].

The association between dietary vitamin C intake and breast cancer survival is inconsistent, and few studies have specifically examined vitamin C supplement use among women with breast cancer. In a meta-analysis study [109] ten identified studies examined vitamin C supplement use ($n = 6$) and dietary vitamin C intake ($n = 7$) and included 17,696 breast cancer cases, 2791 total deaths, and 1558 breast cancer-specific deaths. The summary RR (95% CI) for post-diagnosis vitamin C supplement use was 0.81 (95% CI, 0.72–0.91) for total mortality and 0.85 (95% CI, 0.74–0.99) for breast cancer-specific mortality. The summary RR for a 100 mg per day increase in dietary vitamin C intake was 0.73 (95% CI, 0.59–0.89) for total mortality and 0.78 (95% CI, 0.64–0.94) for breast cancer-specific mortality. It was concluded that post-diagnosis vitamin C supplement use may be associated with a reduced risk of mortality. Dietary vitamin C intake was also statistically significantly associated with a reduced risk of total mortality and breast cancer-specific mortality [109].

Cancer screening has also been studied as factor in survival [110], but this is not the subject of this chapter.

References Cited

1. Siegel, R. L., Miller, K. D., & Jemal, A. (2015). Cancer statistics. *CA: A Cancer Journal for Clinicians, 65*, 5–29.
2. Veronesi, U., Cascinelli, N., Mariani, L., Greco, M., Saccozzi, R., Luini, A., Aguilar, M., & Marubini, E. (2002). Twenty-year follow-up of a randomized study comparing breast-conserving surgery with radical mastectomy for early breast cancer. *The New England Journal of Medicine, 347*, 1227–1232.
3. Blichert-Toft, M., Nielsen, M., During, M., Moller, S., Rank, F., Overgaard, M., & Mouridsen, H. T. (2008). Long-term results of breast conserving surgery vs. mastectomy for early stage invasive breast cancer: 20-year follow-up of the Danish randomized DBCG-82TM protocol. *Acta Oncologica, 47*, 672–681.
4. Nielsen, H. M., Overgaard, M., Grau, C., Jensen, A. R., & Overgaard, J. (2006). Loco-regional recurrence after mastectomy in high-risk breast cancer–risk and prognosis. An analysis of patients from the DBCG 82 b&c randomization trials. *Radiotherapy and Oncology, 79*, 147–155.
5. Early Breast Cancer Trialists' Collaborative Group (EBCTCG). (2005). Effects of chemotherapy and hormonal therapy for early breast cancer on recurrence and 15-year survival: An overview of the randomised trials. *Lancet, 365*, 1687–1717.

6. Cheang, M. C., Voduc, D., Bajdik, C., Leung, S., McKinney, S., Chia, S. K., et al. (2008). Basallike breast cancer defined by five biomarkers has superior prognostic value than triple-negative phenotype. *Clinical Cancer Research, 14*, 1368–1376.
7. Esserman, L. J., Moore, D. H., Tsing, P. J., Chu, P. W., Yau, C., Ozanne, E., et al. (2011). Biologic markers determine both the risk and the timing of recurrence in breast cancer. *Breast Cancer Research and Treatment, 129*, 607–616.
8. Davies, C., Pan, H., Godwin, J., Gray, R., Arriagada, R., Raina, V., et al. (2013). Long-term effects of continuing adjuvant tamoxifen to 10 years versus stopping at 5 years after diagnosis of oestrogen receptor-positive breast cancer: ATLAS, a randomised trial. *Lancet, 381*, 805–816.
9. Jakesz, R., Greil, R., Gnant, M., Schmid, M., Kwasny, W., Kubista, E., et al. (2007). Extended adjuvant therapy with anastrozole among postmenopausal breast cancer patients: Results from the randomized Austrian Breast and Colorectal Cancer Study Group Trial 6a. *Journal of the National Cancer Institute, 99*, 1845–1853.
10. Sestak, I., Dowsett, M., Zabaglo, L., Lopez-Knowles, E., Ferree, S., Cowens, J. W., et al. (2013). Factors predicting late recurrence for estrogen receptor-positive breast cancer. *Journal of the National Cancer Institute, 105*, 1504–1511.
11. Visvader, J. E., & Lindeman, G. J. (2012). Cancer stem cells: Current status and evolving complexities. *Cell Stem Cell, 10*, 717–728.
12. El Helou, R., Wicinski, J., Guille, A., Adelaide, J., Finetti, P., Bertucci, F., Chaffanet, M., Birnbaum, D., Charafe-Jauffret, E., & Ginestier, C. (2014). Brief reports: A distinct DNA methylation signature defines breast cancer stem cells and predicts cancer outcome. *Stem Cells, 32*, 3031–3036.
13. Braun, S., Vogl, F. D., Naume, B., Janni, W., Osborne, M. P., Coombes, R. C., Schlimok, G., Diel, I. J., Gerber, B., Gebauer, G., Pierga, J. Y., Marth, C., Oruzio, D., Wiedswang, G., Solomayer, E. F., Kundt, G., Strobl, B., Fehm, T., Wong, G. Y., Bliss, J., Vincent-Salomon, A., & Pantel, K. (2005). A pooled analysis of bone marrow micrometastasis in breast cancer. *The New England Journal of Medicine, 353*, 793–802.
14. Berman, A. T., Thukral, A. D., Hwang, W. T., Solin, L. J., & Vapiwala, N. (2013). Incidence and patterns of distant metastases for patients with early-stage breast cancer after breast conservation treatment. *Clinical Breast Cancer, 13*, 88–94.
15. Baccelli, I., Schneeweiss, A., Riethdorf, S., Stenzinger, A., Schillert, A., Vogel, V., Klein, C., Saini, M., Bauerle, T., Wallwiener, M., Holland-Letz, T., Hofner, T., Sprick, M., Scharpff, M., Marme, F., Sinn, H. P., Pantel, K., Weichert, W., & Trumpp, A. (2013). Identification of a population of blood circulating tumor cells from breast cancer patients that initiates metastasis in a xenograft assay. *Nature Biotechnology, 31*, 539–544.
16. Aceto, N., Bardia, A., Miyamoto, D. T., Donaldson, M. C., Wittner, B. S., Spencer, J. A., Yu, M., Pely, A., Engstrom, A., Zhu, H., Brannigan, B. W., Kapur, R., Stott, S. L., Shioda, T., Ramaswamy, S., Ting, D. T., Lin, C. P., Toner, M., Haber, D. A., & Maheswaran, S. (2014). Circulating tumor cell clusters are oligoclonal precursors of breast cancer metastasis. *Cell, 158*, 1110–1122.
17. Krebs, M. G., Metcalf, R. L., Carter, L., Brady, G., Blackhall, F. H., & Dive, C. (2014). Molecular analysis of circulating tumor cells-biology and biomarkers. *Nature Reviews. Clinical Oncology, 11*, 129–144.
18. Aggarwal, C., Meropol, N. J., Punt, C. J., Iannotti, N., Saidman, B. H., Sabbath, K. D., Gabrail, N. Y., Picus, J., Morse, M. A., Mitchell, E., Miller, M. C., & Cohen, S. J. (2013). Relationship among circulating tumor cells, CEA and overall survival in patients with metastatic colorectal cancer. *Annals of Oncology, 24*, 420–428.
19. Deneve, E., Riethdorf, S., Ramos, J., Nocca, D., Coffy, A., Daures, J. P., Maudelonde, T., Fabre, J. M., Pantel, K., & Alix-Panabieres, C. (2013). Capture of viable circulating tumor cells in the liver of colorectal cancer patients. *Clinical Chemistry, 59*, 1384–1392.
20. Hou, J. M., Krebs, M. G., Lancashire, L., Sloane, R., Backen, A., Swain, R. K., Priest, L. J., Greystoke, A., Zhou, C., Morris, K., Ward, T., Blackhall, F. H., & Dive, C. (2012). Clinical

significance and molecular characteristics of circulating tumor cells and circulating tumor microemboli in patients with small-cell lung cancer. *Journal of Clinical Oncology, 30*, 525–532.

21. Krebs, M. G., Sloane, R., Priest, L., Lancashire, L., Hou, J. M., Greystoke, A., Ward, T. H., Ferraldeschi, R., Hughes, A., Clack, G., Ranson, M., Dive, C., & Blackhall, F. H. (2011). Evaluation and prognostic significance of circulating tumor cells in patients with non-small-cell lung cancer. *Journal of Clinical Oncology, 29*, 1556–1563.
22. Scher, H. I., Jia, X., de Bono, J. S., Fleisher, M., Pienta, K. J., Raghavan, D., & Heller, G. (2009). Circulating tumour cells as prognostic markers in progressive, castration-resistant prostate cancer: A reanalysis of IMMC38 trial data. *The Lancet Oncology, 10*, 233–239.
23. Zhang, L., Riethdorf, S., Wu, G., Wang, T., Yang, K., Peng, G., Liu, J., & Pantel, K. (2012). Meta-analysis of the prognostic value of circulating tumor cells in breast cancer. *Clinical Cancer Research, 18*, 5701–5710.
24. Alix-Panabieres, C., & Pantel, K. (2014). Challenges in circulating tumour cell research. *Nature Reviews. Cancer, 14*, 623–631.
25. Labelle, M., Begum, S., & Hynes, R. O. (2011). Direct signaling between platelets and cancer cells induces an epithelial–mesenchymal-like transition and promotes metastasis. *Cancer Cell, 20*, 576–590.
26. Yu, M., Bardia, A., Wittner, B. S., Stott, S. L., Smas, M. E., Ting, D. T., Isakoff, S. J., Ciciliano, J. C., Wells, M. N., Shah, A. M., Concannon, K. F., Donaldson, M. C., Sequist, L. V., Brachtel, E., Sgroi, D., Baselga, J., Ramaswamy, S., Toner, M., Haber, D. A., & Maheswaran, S. (2013). Circulating breast tumor cells exhibit dynamic changes in epithelial and mesenchymal composition. *Science, 339*, 580–584.
27. Wels, J., Kaplan, R. N., Rafii, S., & Lyden, D. (2008). Migratory neighbors and distant invaders: Tumor-associated niche cells. *Genes & Development, 22*, 559–574.
28. Psaila, B., & Lyden, D. (2009). The metastatic niche: Adapting the foreign soil. *Nature Reviews. Cancer, 9*, 285–293.
29. Aktary, Z., & Pasdar, M. (2012). Plakoglobin: Role in tumorigenesis and metastasis. *International Journal of Cell Biology, 2012*, 189521.
30. Todorovic, V., Desai, B. V., Patterson, M. J., Amargo, E. V., Dubash, A. D., Yin, T., Jones, J. C., & Green, K. J. (2010). Plakoglobin regulates cell motility through Rho- and fibronectin-dependent Src signaling. *Journal of Cell Science, 123*, 3576–3586.
31. Shair, K. H., Schnegg, C. I., & Raab Traub, N. (2008). EBV latent membrane protein 1 effects on plakoglobin, cell growth, and migration. *Cancer Research, 68*, 6997–7005.
32. Rieger-Christ, K. M., Ng, L., Hanley, R. S., Durrani, O., Ma, H., Yee, A. S., Libertino, J. A., & Summerhayes, I. C. (2005). Restoration of plakoglobin expression in bladder carcinoma cell lines suppresses cell migration and tumorigenic potential. *British Journal of Cancer, 92*, 2153–2159.
33. Yin, T., Getsios, S., Caldelari, R., Kowalczyk, A. P., Muller, E. J., Jones, J. C., & Green, K. J. (2005). Plakoglobin suppresses keratinocyte motility through both cell-cell adhesion-dependent and -independent mechanisms. *Proc Natl Acad Sci USA., 102*, 5420–5425.
34. Kolligs, F. T., Kolligs, B., Hajra, K. M., Hu, G., Tani, M., Cho, K. R., & Fearon, E. R. (2000). Gamma-catenin is regulated by the APC tumor suppressor and its oncogenic activity is distinct from that of beta-catenin. *Genes & Development, 14*, 1319–1331.
35. Hakimelahi, S., Parker, H. R., Gilchrist, A. J., Barry, M., Li, Z., Bleackley, R. C., & Pasdar, M. (2000). Plakoglobin regulates the expression of the anti-apoptotic protein BCL-2. *The Journal of Biological Chemistry, 275*, 10905–10911.
36. Shiina, H., Breault, J. E., Basset, W. W., Enokida, H., Urakami, S., Li, L. C., Okino, S. T., Deguchi, M., Kaneuchi, M., Terashima, M., Yoneda, T., Shigeno, K., Carroll, P. R., Igawa, M., & Dahiya, R. (2005). Functional loss of the gamma-catenin gene through epigenetic and genetic pathways in human prostate cancer. *Cancer Research, 65*, 2130–2138.

37. Lu, L., Zeng, H., Gu, X., & Ma, W. (2015). Circulating tumor cell clusters-associated gene plakoglobin and breast cancer survival. *Breast Cancer Research and Treatment, 151*, 491–500.
38. Al-Hajj, M., Wicha, M. S., Benito-Hernandez, A., Morrison, S. J., & Clarke, M. F. (2003). Prospective identification of tumorigenic breast cancer cells. *Proceedings of the National Academy of Sciences of the United States of America, 100*, 3983–3988.
39. Ginestier, C., Hur, M. H., Charafe-Jauffret, E., Monville, F., Dutcher, J., Brown, M., Jacquemier, J., Viens, P., Kleer, C. G., Liu, S., et al. (2007). ALDH1 is a marker of normal and malignant human mammary stem cells and a predictor of poor clinical outcome. *Cell Stem Cell, 1*, 555–567.
40. Koppaka, V., Thompson, D. C., Chen, Y., Ellermann, M., Nicolaou, K. C., Juvonen, R. O., Petersen, D., Deitrich, R. A., Hurley, T. D., & Vasiliou, V. (2012). Aldehyde dehydrogenase inhibitors: A comprehensive review of the pharmacology, mechanism of action, substrate specificity, and clinical application. *Pharmacological Reviews, 64*, 520–539.
41. Neumeister, V., Agarwal, S., Bordeaux, J., Camp, R. L., & Rimm, D. L. (2010). In situ identification of putative cancer stem cells by multiplexing ALDH1, CD44, and cytokeratin identifies breast cancer patients with poor prognosis. *The American Journal of Pathology, 176*, 2131–2138.
42. Liu, Y., Lv, D. L., Duan, J. J., Xu, S. L., Zhang, J. F., Yang, X. J., Zhang, X., Cui, Y. H., Bian, X. W., & Yu, S. C. (2014). ALDH1A1 expression correlates with clinicopathologic features and poor prognosis of breast cancer patients: A systematic review and meta-analysis. *BMC Cancer, 14*, 444–454.
43. Tierney, J. F., Stewart, L. A., Ghersi, D., Burdett, S., & Sydes, M. R. (2007). Practical methods for incorporating summary time-to-event data into meta-analysis. *Trials, 8*, 16–26.
44. Sakakibara, M., Fujimori, T., Miyoshi, T., Nagashima, T., Fujimoto, H., Suzuki, H. T., Ohki, Y., Fushimi, K., Yokomizo, J., Nakatani, Y., & Miyazaki, M. (2012). Aldehyde dehydrogenase 1-positive cells in axillary lymph node metastases after chemotherapy as a prognostic factor in patients with lymph node-positive breast cancer. *Cancer, 118*, 3899–3910.
45. Yu, P., Zhou, L. W. J., Jiang, A. F., & Li, K. (2010). Prognostic relevance of ALDH1 in breast cancer: A clinicopathological study of 96 cases. *The Chinese-German Journal of Clinical Oncology, 9*, 31–35.
46. Charafe-Jauffret, E., Ginestier, C., Iovino, F., Tarpin, C., Diebel, M., Esterni, B., Houvenaeghel, G., Extra, J. M., Bertucci, F., Jacquemier, J., Xerri, L., Dontu, G., Stassi, G., Xiao, Y., Barsky, S. H., Birnbaum, D., Viens, P., & Wicha, M. S. (2010). Aldehyde dehydrogenase 1-positive cancer stem cells mediate metastasis and poor clinical outcome in inflammatory breast cancer. *Clinical Cancer Research, 16*, 45–55.
47. Dong, Y., Bi, L. R., Xu, N., Yang, H. M., Zhang, H. T., Ding, Y., Shi, A. P., & Fan, Z. M. (2013). The expression of aldehyde dehydrogenase 1 in invasive primary breast tumors and axillary lymph node metastases is associated with poor clinical prognosis. *Pathology, Research and Practice, 209*, 555–561.
48. Lee, H. E., Kim, J. H., Kim, Y. J., Choi, S. Y., Kim, S. W., Kang, E., Chung, I. Y., Kim, I. A., Kim, E. J., Choi, Y., Ryu, H. S., & Park, S. Y. (2011). An increase in cancer stem cell population after primary systemic therapy is a poor prognostic factor in breast cancer. *British Journal of Cancer, 104*, 1730–1738.
49. Morimoto, K., Kim, S. J., Tanei, T., Shimazu, K., Tanji, Y., Taguchi, T., Tamaki, Y., Terada, N., & Noguchi, S. (2009). Stem cell marker aldehyde dehydrogenase 1-positive breast cancers are characterized by negative estrogen receptor, positive human epidermal growth factor receptor type 2, and high Ki67 expression. *Cancer Science, 100*, 1062–1068.
50. Nalwoga, H., Arnes, J. B., Wabinga, H., & Akslen, L. A. (2010). Expression of aldehyde dehydrogenase 1 (ALDH1) is associated with basal-like markers and features of aggressive tumours in African breast cancer. *British Journal of Cancer, 102*, 369–375.

51. Resetkova, E., Reis-Filho, J. S., Jain, R. K., Mehta, R., Thorat, M. A., Nakshatri, H., & Badve, S. (2010). Prognostic impact of ALDH1 in breast cancer: A story of stem cells and tumor microenvironment. *Breast Cancer Research and Treatment, 123*, 97–108.
52. Tan, E. Y., Thike, A. A., & Tan, P. H. (2013). ALDH1 expression is enriched in breast cancers arising in young women but does not predict outcome. *British Journal of Cancer, 109*, 109–113.
53. Yasuyo, O., Umekita, Y., Yoshioka, T., Souda, M., Rai, Y., Sagara, Y., Sagara, Y., Sagara, Y., & Tanimoto, A. (2011). Aldehyde dehydrogenase 1 expression predicts poor prognosis in triple-negative breast cancer. *Histopathology, 59*, 776–780.
54. Yoshioka, T., Umekita, Y., Ohi, Y., Souda, M., Sagara, Y., Rai, Y., & Tanimoto, A. (2011). Aldehyde dehydrogenase 1 expression is a predictor of poor prognosis in node-positive breast cancers: A long-term follow-up study. *Histopathology, 58*, 608–616.
55. Neve, R. M., Chin, K., Fridlyand, J., Yeh, J., Baehner, F. L., Fevr, T., Clark, L., Bayani, N., Coppe, J. P., Tong, F., Speed, T., Spellman, P. T., DeVries, S., Lapuk, A., Wang, N. J., Kuo, W. L., Stilwell, J. L., Pinkel, D., Albertson, D. G., Waldman, F. M., McCormick, F., Dickson, R. B., Johnson, M. D., Lippman, M., Ethier, S., Gazdar, A., & Gray, J. W. (2006). A collection of breast cancer cell lines for the study of functionally distinct cancer subtypes. *Cancer Cell, 10*, 515–527.
56. Teicher, B. A., & Fricker, S. P. (2010). CXCL12 (SDF-1)/CXCR4 pathway in cancer. *Clinical Cancer Research, 16*, 2927–2931.
57. John, M., & Busillo, J. L. B. (1768). Regulation of CXCR4 Signaling. *Biochimica et Biophysica Acta, 2007*, 952–963.
58. Kucia, M., Reca, R., Miekus, K., Wanzeck, J., Wojakowski, W., Janowska-Wieczorek, A., Ratajczak, J., & Ratajczak, M. Z. (2005). Trafficking of normal stem cells and metastasis of cancer stem cells involve similar mechanisms: Pivotal role of the SDF-1-CXCR4 axis. *Stem Cells, 23*, 879–894.
59. Zlotnik, A. (2008). New insights on the role of CXCR4 in cancer metastasis. *The Journal of Pathology, 215*, 211–213.
60. Luker, K. E., & Luker, G. D. (2006). Functions of CXCL12 and CXCR4 in breast cancer. *Cancer Letters, 238*, 30–41.
61. Zhang, Z., Ni, C., Chen, W., Wu, P., Wang, Z., Yin, J., Huang, J., & Qiu, F. (2014). Expression of CXCR4 and breast cancer prognosis: A systematic review and meta-analysis. *BMC Cancer, 14*, 49–57.
62. Liles, W. C., Broxmeyer, H. E., Rodger, E., Wood, B., Hubel, K., Cooper, S., Hangoc, G., Bridger, G. J., Henson, G. W., Calandra, G., et al. (2003). Mobilization of hematopoietic progenitor cells in healthy volunteers by AMD3100, a CXCR4 antagonist. *Blood, 102*, 2728–2730.
63. Gil, M., Seshadri, M., Komorowski, M. P., Abrams, S. I., & Kozbor, D. (2013). Targeting CXCL12/CXCR4 signaling with oncolytic virotherapy disrupts tumor vasculature and inhibits breast cancer metastases. *Proceedings National Academy of Sciences United States of America, 110*, E1291–E1300.
64. Domanska, U. M., Kruizinga, R. C., Nagengast, W. B., Timmer-Bosscha, H., Huls, G., de Vries, E. G., & Walenkamp, A. M. (2013). A review on CXCR4/CXCL12 axis in oncology: No place to hide. *European Journal of Cancer, 49*, 219–230.
65. Balkwill, F. (2004). Cancer and the chemokine network. *Nature Reviews. Cancer, 4*, 540–550.
66. Ferlay, J., Soerjomataram, I., Dikshit, R., et al. (2015). Cancer incidence and mortality worldwide: Sources, methods, and major patterns in GLOBOCAN 2012. *International Journal of Cancer, 136*, E359–E386.
67. Early Breast Cancer Trialists' Collaborative Group. (EBCTCG). (2012). Comparisons between difference polychemotherapy regimens for early breast cancer: Meta-analyses of long-term outcome among 100,000 women in 123 randomized trials. *Lancet, 379*, 432–444.
68. National comprehensive care network. NCCN clinical practice guidelines in oncology. Breast cancer (Version 1. 2016). http://www.nccn.org/professionals/physician_gls/pdf/breast.pdf

69. Ng, W., Delaney, G. P., Jacob, S., & Barton, M. B. (2010). Estimation of an optimal chemotherapy utilisation rate for breast cancer: Setting an evidenced-based benchmark for the best-quality cancer care. *European Journal of Cancer, 46*, 703–712.
70. Raphael, M. J., Biagi, J. J., Kong, W., Mates, M., Booth, C. M., & Mackillop, W. J. (2016). The relationship between time to initiation of adjuvant chemotherapy and survival in breast cancer: A systematic review and meta-analysis. *Breast Cancer Research and Treatment, 160*, 17–28.
71. Schabel, F. M. (1977). Rationale for adjuvant chemotherapy. *Cancer, 39*(6 Suppl), 2875–2882.
72. Schatten, W. E. (1958). An experimental study of postoperative tumor metastases. I. Growth of pulmonary metastases following total removal of primary leg tumor. *Cancer, 11*, 455–459.
73. Gunduz, N., Fisher, B., & Saffer, E. A. (1979). Effect of surgical removal on the growth and kinetics of residual tumor. *Cancer Research, 39*, 3861–3865.
74. Yu, K. D., Huang, S., Zhang, J. X., Liu, G. Y., & Shao, Z. M. (2013). Association between delayed initiation of adjuvant CMF or anthracycline-based chemotherapy and survival in breast cancer: A systematic review and meta-analysis. *BMC Cancer, 13*, 240–251.
75. Cedolini, C., Bertozzi, S., Londero, A. P., Bernardi, S., Seriau, L., Concina, S., Cattin, F., & Risaliti, A. (2014). Type of breast cancer diagnosis, screening, and survival. *Clinical Breast Cancer, 14*, 235–240.
76. Norton, L. (2005). Conceptual and practical implications of breast tissue geometry: Toward a more effective, less toxic therapy. *The Oncologist, 10*, 370–381.
77. Petrelli, F., Cabiddu, M., Coinu, A., Borgonovo, K., Ghilardi, M., Lonati, V., & Barni, S. (2015). Adjuvant dose-dense chemotherapy in breast cancer: A systematic review and meta-analysis of randomized trials. *Breast Cancer Research and Treatment, 151*, 251–259.
78. Moynahan, M. E., Chiu, J. W., Koller, B. H., & Jasin, M. (1999). Brca1 controls homology-directed DNA repair. *Molecular Cell, 94*, 511–518.
79. Orban, T. I., & Olah, E. (2003). Emerging roles of BRCA1 alternative splicing. *Molecular Pathology, 56*, 191–197.
80. Huzarski, T., Byrski, T., Gronwald, J., Gorski, B., Domagala, P., Cybulski, C., et al. (2013). Ten-year survival in patients with BRCA1-negative and BRCA1-positive breast cancer. *Journal of Clinical Oncology, 31*, 3191–3196.
81. Foulkes, W. D., Chappuis, P. O., Wong, N., Brunet, J. S., Vesprini, D., Rozen, F., et al. (2000). Primary node negative breast cancer in BRCA1 mutation carriers has a poor outcome. *Annals of Oncology, 11*, 307–313.
82. Moller, P., Evans, D. G., Reis, M. M., Gregory, H., Anderson, E., Maehle, L., et al. (2007). Surveillance for familial breast cancer: Differences in outcome according to BRCA mutation status. *International Journal of Cancer, 121*, 1017–1020.
83. Lee, E. H., Park, S. K., Park, B., Kim, S. W., Lee, M. H., Ahn, S. H., et al. (2010). Effect of BRCA1/2 mutation on short-term and long-term breast cancer survival: A systematic review and meta-analysis. *Breast Cancer Research and Treatment, 122*, 11–25.
84. Goodwin, P. J., Phillips, K. A., West, D. W., Ennis, M., Hopper, J. L., John, E. M., et al. (2012). Breast cancer prognosis in BRCA1 and BRCA2 mutation carriers. An International Prospective Breast Cancer Family Registry population-based cohort study. *Journal of Clinical Oncology, 30*, 19–26.
85. Cortesi, L., Masini, C., Cirilli, C., Medici, V., Marchi, I., Cavazzini, G., et al. (2010). Favourable ten-year overall survival in a Caucasian population with high probability of hereditary breast cancer. *BMC Cancer, 10*, 90–96.
86. Tryggvadottir, L., Olafsdottir, E. J., Olafsdottir, G. H., Sigurdsson, H., Johannsson, O. T., Bjorgvinsson, E., et al. (2013). Tumour diploidy and survival in breast cancer patients with BRCA2 mutations. *Breast Cancer Research and Treatment, 140*, 375–384.
87. Zhong, Q., Peng, H. L., Zhao, X., Zhang, L., & Hwang, W. T. (2015). Effects of BRCA1- and BRCA2-related mutations on ovarian and breast cancer survival: A meta-analysis. *Clinical Cancer Research, 21*, 211–220.

88. Nicoletto, M. O., Donach, M., De Nicolo, A., Artioli, G., Banna, G., & Monfardini, S. (2001). BRCA-1 and BRCA-2 mutations as prognostic factors in clinical practice and genetic counselling. *Cancer Treatment Reviews, 27*, 295–304.
89. Robson, M. E. (2002). Clinical considerations in the management of individuals at risk for hereditary breast and ovarian cancer. *Cancer Control, 9*, 457–465.
90. Honrado, E., Osorio, A., Palacios, J., & Benitez, J. (2006). Pathology and gene expression of hereditary breast tumors associated with BRCA1, BRCA2 and CHEK2 gene mutations. *Oncogene, 25*, 5837–5845.
91. Phillips, K. A. (2001). Current perspectives on BRCA1- and BRCA2-associated breast cancers. *Internal Medicine Journal, 31*, 349–356.
92. Narod, S. A., & Foulkes, W. D. (2004). BRCA1 and BRCA2: 1994 and beyond. *Nature Reviews. Cancer, 4*, 665–676.
93. van den Broek, A. J., Schmidt, M. K., van 't Veer, L. J., Tollenaar, R. A., & van Leeuwen, F. E. (2015). Worse breast cancer prognosis of BRCA1/BRCA2 mutation carriers: What's the evidence? A systematic review with meta-analysis. *PLoS One, 10*, e0120189.
94. Valachis, A., Nearchou, A. D., & Lind, P. (2014). Surgical management of breast cancer in BRCA-mutation carriers: A systematic review and meta-analysis. *Breast Cancer Research and Treatment, 144*, 443–455.
95. Tang, Y., Zhou, X., Ji, J., Chen, L., Cao, J., Luo, J., & Zhang, S. (2015). High expression levels of miR-21 and miR-210 predict unfavorable survival in breast cancer: A systemic review and meta-analysis. *The International Journal of Biological Markers, 30*, e347–e358.
96. Pan, F., Mao, H., Deng, L., Li, G., & Geng, P. (2014). Prognostic and clinicopathological significance of microRNA-21 overexpression in breast cancer: A meta-analysis. *International Journal of Clinical and Experimental Pathology, 7*, 5622–5633.
97. Pavlou, M. P., Dimitromanolakis, A., Martinez-Morillo, E., Smid, M., Foekens, J. A., & Diamandis, E. P. (2014). Integrating meta-analysis of microarray data and targeted proteomics for biomarker identification: Application in breast cancer. *Journal of Proteome Research, 13*, 2897–2909.
98. Abravanel, D. L., Belka, G. K., Pan, T. C., Pant, D. K., Collins, M. A., Sterner, C. J., & Chodosh, L. A. (2015). Notch promotes recurrence of dormant tumor cells following HER2/neu-targeted therapy. *The Journal of Clinical Investigation, 125*, 2484–2496.
99. Yuan, X., Zhang, M., Wu, H., Xu, H., Han, N., Chu, Q., Yu, S., Chen, Y., & Wu, K. (2015). Expression of Notch1 correlates with breast cancer progression and prognosis. *PLoS One, 10*, e0131689.
100. Yan, S., Jiao, X., Zou, H., & Li, K. (2015). Prognostic significance of c-Met in breast cancer: A meta-analysis of 6010 cases. *Diagnostic Pathology, 10*, 62–71.
101. Lin, S., Gan, Z., Han, K., Yao, Y., & Min, D. (2015). Interleukin-6 as a prognostic marker for breast cancer: A meta-analysis. *Tumori, 101*, 535–541.
102. Kim, H. J., Kim, A., Ahn, H., Ahn, I. M., Choi, J., & Chang, H. (2015). Meta-analysis reveals no significant correlation between breast cancer survival and ErbB3 expression. *APMIS, 123*, 383–393.
103. Li, Y., Ma, X., Wu, X., Liu, X., & Liu, L. (2014). Prognostic significance of survivin in breast cancer: Meta-analysis. *The Breast Journal, 20*, 514–524.
104. Lahart, I. M., Metsios, G. S., Nevill, A. M., & Carmichael, A. R. (2015). Physical activity, risk of death and recurrence in breast cancer survivors: A systematic review and meta-analysis of epidemiological studies. *Acta Oncologica, 54*, 635–654.
105. Bérubé, S., Lemieux, J., Moore, L., Maunsell, E., & Brisson, J. (2014). Smoking at time of diagnosis and breast cancer-specific survival: New findings and systematic review with meta-analysis. *Breast Cancer Research, 16*(2), R42.
106. Rose, A. A., Elser, C., Ennis, M., & Goodwin, P. J. (2013). Blood levels of vitamin D and early stage breast cancer prognosis: A systematic review and meta-analysis. *Breast Cancer Research and Treatment, 141*, 331–339.

107. Kim, Y., & Je, Y. (2014). Vitamin D intake, blood 25(OH)D levels, and breast cancer risk or mortality: A meta-analysis. *British Journal of Cancer, 110*, 2772–2784.
108. Mohr, S. B., Gorham, E. D., Kim, J., Hofflich, H., & Garland, C. F. (2014). Meta-analysis of vitamin D sufficiency for improving survival of patients with breast cancer. *Anticancer Research, 34*, 1163–1166.
109. Harris, H. R., Orsini, N., & Wolk, A. (2014). Vitamin C and survival among women with breast cancer: A meta-analysis. *European Journal of Cancer, 50*, 1223–1231.
110. Hanley, J. A., McGregor, M., Liu, Z., Strumpf, E. C., & Dendukuri, N. (2013). Measuring the mortality impact of breast cancer screening. *Canadian Journal of Public Health, 104*, e437–e442.

Chapter 3
Synopsis of the Treatment of Breast Cancer

Abstract The author emphasizes the importance and the wisdom of the St. Gallen Consensus Conference that is a body of internationally recognized specialists that has developed consensus on the treatment of this breast cancer. Primary tumor resection is a well-established protocol. However, a topic of discussion has been which is the acceptable margin of reception? The agreement is that the "minimum acceptable surgical margin is 'no ink on invasive tumor', and 1–2 mm' clearance." Sentinel node (SN) biopsy must be considered as the standard of care. The use of radiation therapy has allowed women with early stage disease to maintain integrity of their body and those with advanced disease improve their quality of life. *After breast-conserving surgery, radiation should include breast only in the case of a negative SN/axilla.* The standard evaluation of breast cancer for clinical purposes involves IHC characterization of estrogen receptors (ER) and progesterone receptor (PR) status. Hormone receptor-positive breast cancers account for around 75–80% of all cases, and standardized IHC assays for the routine testing of ER and PR are used to guide the selection of patients for hormonal-based therapies. In addition to the use of SERMS, it has been recently emphasized the importance of ovarian function suppression (OFS) as adjuvant treatment option in premenopausal patients. The use of chemotherapy has slowly improved cancer recurrence, mainly doxorubicin and cyclophosphamide–methotrexate–fluorouracil (CMF) regimens, with a benefit of about 4%, and by the use of anthracyclines with additional benefit of another 4%. The use of taxanes added another 5% improvement. The two drugs that summarize the achievement of molecular target specific and immunotherapy are trastuzumab (Herceptin) and pertuzumab (Perjeta). Both are antibody-based agents that target the signaling protein HER2, which is produced in abundance in 20–25% of breast tumors. In summary the rate of HER2 positivity in breast cancer is about 15%. In HER2+ early breast cancer, neo-adjuvant therapy using dual antibody-based HER2 blockade together with taxane-based chemotherapy is now standard in stages II–III. In HER2+ early breast cancer, the APHINITY trial showed that dual antibody-based HER2 blockade together with adjuvant chemotherapy significantly prolongs invasive disease-free survival. And in HER2+ advanced breast cancer, T-DM1 significantly prolongs overall survival beyond first line despite substantial cross-over rates in the phase III trials.

Keywords St. Gallen Consensus Conference · Primary tumor resection · Minimum *acceptable surgical margin* · Sentinel node (SN) biopsy · Estrogen

J. Russo, *The Future of Prevention and Treatment of Breast Cancer*,
https://doi.org/10.1007/978-3-030-72815-1_3

receptors (ER) · Progesterone receptor (PR) status · SERMS · Doxorubicin · Cyclophosphamide–methotrexate–fluorouracil (CMF) · Trastuzumab (Herceptin) · Pertuzumab (Perjeta)

3.1 Current Treatment Strategies

This is not a treatise but a brief summary of the present status of breast cancer treatment, and the author will emphasize the importance and the wisdom of the St. Gallen Consensus Conference that is a body of internationally recognized specialists that has developed consensus on the treatment of this disease [1]. Therefore, in this chapter on the treatment of breast cancer, the author will summarize the general guidelines used by this international body making only emphasis in those areas that are still controversial and that need further studies. This chapter is the basic background for understanding the topics discussed in future chapters.

3.2 Surgery

Primary tumor resection is a well-established protocol [2]; however a topic of discussion has been which is the acceptable margin of reception? It is an agreement that the "minimum acceptable surgical margin is 'no ink on invasive tumor', and 1–2 mm' clearance. The margin required will not depend on tumor biology and should not be greater when age is less than 40 years or for lobular histology and should not ne be greater after neo-adjuvant chemotherapy. The margins should not be greater in the presence of extensive intraductal component and no greater for pure ductal carcinoma in situ than for invasive disease" [1].

Sentinel node (SN) biopsy must be considered as the standard of care [3]. The technique of sentinel lymph node biopsy for the evaluation and management of breast carcinoma [4–9] is based on the concept that if the sentinel node is negative, the other nodes of that group will also be negative in nearly all instances, whereas if it is positive, the chance that there will be additional metastases in that nodal group is about one third. The study of the sentinel node that proves negative on frozen section should include at least three step sections stained with H&E plus and at least one section immunostained for keratin [10–13]. The immunostain of choice is a keratin cocktail, such as AEl/AE3 [14]. A cluster of metastatic cells in sentinel lymph node is highlighted with keratin stain. Diagnosis of metastatic carcinoma on the basis of keratin-positive cells that is not evident in the H&E preparations needs to be very carefully considered. SN is advised after neo-adjuvant chemotherapy and successful down staging of an initially clinically positive axillary node. When the SN turns out to be positive, it is recommended to be followed by axillary dissection [1].

3.3 Radiotherapy

The use of radiation therapy has allowed women with early stage disease to maintain integrity of their body and those with advanced disease improve their quality of life. This has been supported by several randomized trials with long-term follow-up showing similar results between mastectomy and breast-conserving surgery with radiation [15, 16] . It has also been reported [17] that in women with higher risk or locally advanced disease, radiation also decreases the rate of recurrence by 70–80% following mastectomy [17–20]. Radiation therapy is also recognized as one of the primary modalities to treat complications of metastatic breast cancer such as bone pain, spinal cord compression, or neurologic symptoms due to brain metastasis [21].

The St. Gallen recommendations are "hypofractionated breast irradiation after breast-conserving surgery in patients aged 50+ without chemotherapy or axillary involvement, but also for patients younger than 50 years with uncertainty about patients with prior chemotherapy or axillary lymph node involvement."

"After breast-conserving surgery, radiation should include breast only in the case of a negative SN/axilla. For positive nodes and breast conservation, breast and regional nodes but not internal mammary nodes (IMN) should be irradiated [1].

Post-mastectomy radiotherapy (PMRT) should be standard of care for T3, N+ 1–3 with adverse pathology, but not for all N+ 1–3 nor N+ 1–3 at age <40 years. PMRT should be given to patients with positive SN but no axillary dissection (AD), but not to pN0 patients after AD and <8 nodes unless micro-metastases were found in the latter situation."

"PMRT should include chest wall and regional nodes but not IMN. Radiotherapy after immediate reconstruction should include the reconstructed breast and nodes in most cases, with uncertainty about 'only for tumors with adverse pathological features'.

In general, after neo-adjuvant therapy, the radiotherapy strategies should follow the stage before neo-adjuvant therapy rather than after neo-adjuvant therapy."

3.4 Endocrine Therapy

The standard evaluation of breast cancer for clinical purposes involves IHC characterization of estrogen receptors (ER) and progesterone receptor (PR) status. Hormone receptor-positive breast cancers account for around 75–80% of all cases, and standardized IHC assays for the routine testing of ER and PR are used to guide the selection of patients for hormonal-based therapies. More than 50% of breast cancers have an abundance of estrogen receptors, particularly ER-alpha, the presence of which correlates with a positive prognosis for the patient. The absence of an overexpression of ER-alpha has the inverse prognosis [22]. The presence of estrogen and progesterone receptors in the tumor tissue correlates well with response to hormone therapy and chemotherapy [23, 24]. It is presently accepted the immune-histochemical method that has the versatility to be performed in minute amount of fixed tissue. The two parameters evaluated in immune-histochemical preparations of hormone receptors are the number of tumor cell nuclei stained and the intensity of the reaction. The first is expressed as a percentage of the entire tumor cell nuclei population, and the second is graded as negative, weak, moderate, and strong. The two parameters are sometimes combined into a scoring system, of which three major versions exist [24, 25]. The receptors are 8.5 ER and PR as Biomarkers of Prognosis expressed in both ductal and lobular type of carcinomas; however medullary carcinomas and intraductal carcinomas of the comedocarcinoma type are negative, whereas mucinous carcinomas have the highest rates of positivity.

In addition to the use of SERMS, it has been recently emphasized the importance of ovarian function suppression (OFS) as adjuvant treatment option in premenopausal patients [26, 27]. As a guideline from St. Gallen, it is recommended that for a "premenopausal woman, intermediate grade (G2), node-negative, steroid hormone receptor-positive breast cancer, having not received adjuvant chemotherapy, treatment with tamoxifen alone is the choice. For premenopausal after adjuvant chemotherapy for a pT1, grade 3, node-positive steroid hormone receptor-positive breast cancer it is advisable OFS and exemestane or OFS and tamoxifen" [1].

Recommendations of the indication of OFS must be considered at age $\leq$*35 years, premenopausal estrogen levels after chemotherapy, and high tumor load (4 or more axillary lymph nodes) as arguments in favor of OFS. The preference of OFS plus an aromatase inhibitor (AI) rather than OFS and tamoxifen is in discussion [1]. The strongest arguments in favor of using AI were grade 3 and high tumor load (4 or more axillary lymph nodes), followed by adverse result of multi-gene test and age* $\leq$*35 years. The major consensus is the combination of AI with OFS,* versus *tamoxifen plus OFS as treatment of choice.*

Tamoxifen is considered as adequate treatment in some postmenopausal patients; however, clear factors arguing for inclusion of an AI were high tumor load defined by 4 or more tumor-infiltrated axillary lymph nodes, or unfavorable biology defined by grade 3 or high Ki-67, or HER2 positivity. According to the St. Gallen Consensus "in patients at higher risk, upfront AI was clearly regarded as the preferred mode of therapy and can be switched to tamoxifen after 2 years."

Prolongation of the adjuvant endocrine treatment exceeding the traditional duration of 5 years, and for those postmenopausal patients who were premenopausal at baseline, it is recommended based on several clinical trials [28–30].

3.5 Chemotherapy

The use chemotherapy has slowly improved cancer recurrence, mainly doxorubicin and cyclophosphamide–methotrexate–fluorouracil (CMF) regimens, with a benefit of about 4%, and by the use of anthracyclines with additional benefit of another 4%. The use of taxanes added another 5% improvement, and finally the use of trastuzumab with a 15–20% improvement in outcomes in patients with early breast cancer.

Presently, adjuvant chemotherapy is recommended in cases with histological grade 3 tumor, any positive node; 4 or more positive nodes; high Ki-67; age <35 years; extensive lymphovascular invasion; and low hormone receptor staining.

For the luminal A phenotype that are less responsive to chemotherapy, its use is indicated depending on the tumor size; lymphovascular invasion; 1–3 involved nodes; and ≥4 involved nodes. *In the luminal B*-like tumors, chemotherapy should be recommended only in patients with other indicators of increased risk such as low Oncotype Dx® (Genomic Health, Inc. Redwood City, CA, USA) score but not an intermediate score. Chemotherapy may also be omitted in patients with a low-risk MammaPrint result, a low PAM50 risk of recurrence score, and a low-risk Endo-Predict® (Sividon, Köln, Germany) result [1]. *In luminal B (HER2-negative)* tumors, chemotherapy should include an anthracycline and a taxane. *In TNBC*, chemotherapy should contain anthracyclines and taxanes. Platinum-based regimen is advised only in known BRCA mutation carriers.

In HER2-positive disease, chemotherapy should always be given to patients with stage 2 disease who require anti-HER2 therapy. This chemotherapy should contain anthracyclines and taxanes [1].

Adjuvant bisphosphonates, such as zoledronic acid every 6 months or oral clodronate, are indicated during adjuvant endocrine therapy for postmenopausal patients with endocrine responsive disease.

3.6 Immunotherapy

The two drugs that summarize the achievement of molecular target specific and immunotherapy are trastuzumab (Herceptin) and pertuzumab (Perjeta). Both are antibody-based agents that target the signaling protein HER2, which is produced in abundance in 20–25% of breast tumors. The high levels result in poorly controlled cell growth and proliferation. Patients with HER2-positive tumors are essentially having their cancer eradicated by receiving a double-hit of targeted therapy before surgery. Even patients diagnosed with late-stage, metastatic disease have longer survival rate than those that did not receive the drugs. These two drugs also represent personalized treatment based on specific set of mutations, rather than conventional chemotherapy, which is broadly toxic to healthy as well as dangerous cells. However, cancer returns in many people who receive prompt treatment.

Trastuzumab was the first HER2-targeting drug to reach the market, following a phase 3 clinical trial that showed that the drug improved the odds of survival by 20% in women with metastatic HER2-positive breast cancer [31]. Subsequent data showed that giving people the drug after the surgical removal of early-stage tumors cut the risk of the cancer returning in half [32].

In the US National Cancer Database, 298,937 invasive breast cancers with known HER2 status were diagnosed between 2010 and 2011. HER2-positivity was 14.5%.

There was a difference in HER2-positivity according to geographical region in the USA and patient ethnicity with the lowest incidence in non-Hispanic Whites. Young women had a higher incidence. HER2+ tumors in general had features of higher tumor aggressiveness regarding tumor size, nodal involvement, grade, and lymphovascular invasion. They were also more common in inflammatory breast cancer and Paget's disease with invasion [33].

In 2017, the St. Gallen Consensus recommended that the neo-adjuvant approach should be standard in stage II–III HER2+ tumors [34, 35]. So far, dual HER2 blockade with trastuzumab and pertuzumab is approved for the neo-adjuvant setting only as the NeoSPHERE trial showed that it significantly increases pathological complete response (pCR) together with docetaxel compared to docetaxel and trastuzumab alone. After surgery, the current standard is to complete 1 year of trastuzumab [34, 35].

The APHINITY trial involving 4805 women demonstrated efficacy of adjuvant dual HER blockade in the adjuvant setting. Adding pertuzumab to trastuzumab in the adjuvant setting significantly improved disease-free survival with a small absolute difference in recurrences of 7.1 versus 8.7%. The corresponding estimated 3-year invasive DFS rates were 94.1 versus 93.2%. The absolute differences were somewhat larger in the subgroup of node-positive disease with 3-year iDFS rates of 92.0 versus 90.2%.

According to the St. Gallen Consensus, "the APHINITY trial demonstrates how good the prognosis of patients with HER2+ EBC has become, even with standard therapy. Given the short follow up of the trial, the small absolute differences between the treatment arms, and the change towards neoadjuvant management of HER2þ EBC. Approval of dual blockade based on the trial results will certainly provide more flexibility in treating HER2+ EBC" [36].

The WSG ADAPT umbrella trial that is the use of adjuvant dynamic marker-adjusted personalized therapy with its two HER2 subtrials in HR-positive and HR-negative (HR-) disease is the first trial specifically providing different treatment strategies according to HR status in HER2+ EBC. ADAPT aims at identifying early response markers predicting pCR in a 12-week neo-adjuvant setting and at avoiding overtreatment in EBC. In the ADAPT HER2+/HR- subtrial [37], 92 patients were randomized to 12 weeks trastuzumab and pertuzumab alone versus paclitaxel plus dual antibody blockade. pCR in the chemotherapy containing arm was 90.5 versus 36.3% in the antibody alone arm; it was also higher than pCR in the early responders to antibody alone therapy. More than half of the patients were initially cN0. Thus, paclitaxel plus dual blockade is a highly effective neo-adjuvant regimen in low to intermediate clinical risk patients with HER2+ HR- disease. In the ADAPT HR+/HER2+ subtrial [38], 375 patients with HER2+/HR+ disease were randomized to 12 weeks of T-DM1 (+ endocrine therapy) versus trastuzumab plus endocrine therapy. There was a substantial pCR rate of about 41% in both T-DM1 arms compared to only 15.1% in the trastuzumab based arm. Endocrine therapy did not impact T-DM1 efficacy. Both ADAPT subtrials provide interesting de-escalation strategies for HER2+ EBC that need to be investigated further. The data for paclitaxel plus dual blockade already support this regimen for clinical routine use, in particular once outcome data are available. As demonstrated by ADAPT, neoadjuvant therapy is a good setting for trials exploring further tumor biology-based treatment de-escalation in HER2+ EBC.

The neo-adjuvant PAMELA trial treated 151 patients with HER2+ EBC with lapatinib and trastuzumab for 18 weeks and endocrine therapy, in case of a HR+ tumor. There was a 30% pCR rate (in breast); among patients with HER2-enriched tumors by molecular subtyping, 41% had a pCR compared to only 10% of those patients with other tumor subtypes. The overall percentage of HER2-enriched tumors at baseline was 67%, with an additional 15% of tumors being luminal A, 11% luminal B, 6% basal-like, and 2% normal-like [39]. These data suggest that molecular subtyping may play a future role even in HER2+ disease. In PAMELA, tumor infiltrating lymphocytes (TILs) at day 15 but not at baseline were correlated with pCR. Moreover, combining the TIL count and tumor cellularity at day 15 was an independently predictive of response [40, 41]. Together with the ADAPT results on early response after 2 weeks, these data show that response to anti-HER2 therapy can be clinically seen already after the first cycle of therapy.

The neo-adjuvant GeparSepto trial comparing nab-paclitaxel with paclitaxel in a weekly regimen followed by four cycles of EC included 396 patients with HER2+ EBC who all received dual antibody blockade. pCR rates in the HER2+ cohort were higher than in HER2- disease. Rates were highest in the HER2+/HR- than in the HER2+/HR+ cohort and higher with nab-paclitaxel than with paclitaxel. There were more grade 3 side effects in the HER2+ versus HER2- cohort with grade 3–4 diarrhea in 7.6 vs. 0.9% and febrile neutropenia in 6.3 versus 3.3%. Cardiac side effects were not common. These data show that the GeparSepto regimens are highly effective in HER2+ EBC, yet proper attention needs to be paid to proactive side effect management in patients with chemotherapy plus dual HER2 blockade [40, 41]. A systematic review regarding concurrent administration of radiotherapy and anti-HER2 therapies (lapatinib, trastuzumab, pertuzumab, T-DM1) did not detect any serious adverse events for this treatment concept. The authors, however, underlined the heterogeneity of the existing literature and the necessity for further data and follow-up [42].

MANTICORE 101 or Multidisciplinary Approach to Novel Therapies in Cardio-Oncology Research (MANTICORE 101-Breast) is a Randomized Trial for the Prevention of Trastuzumab-Associated Cardiotoxicity. This was a randomized double-blind, placebo-controlled trial in EBC evaluating the impact of an angiotensin-converting enzyme inhibitor (perindopril) and a b-blocker (bisoprolol) on prevention of trastuzumab-related cardiotoxicity. Both drugs were well tolerated and protected patients who received trastuzumab in EBC against cancer-therapy-related LVEF declines. Nevertheless, they were not able to prevent trastuzumab-mediated left ventricular remodeling which was the primary endpoint of the trial [43]. These results together with the low but still relevant cardiac toxicity figures from the adjuvant trials emphasize the necessity to pay attention to additional potentially cardiotoxic drugs such as anthracyclines in HER2+ EBC and try to avoid them wherever possible.

In the Advance Breast Cancer (ABC) consensus meeting 2017 in Lisbon, it was concluded that based on the substantial OS benefit in the CLEOPATRA trial, the first line standard in HER2+ ABC is taxane chemotherapy together with dual antibody-based HER2 blockade with trastuzumab and pertuzumab. The phase III MARIANNE trial compared trastuzumab plus a taxane with T-DM1 and T-DM1 plus pertuzumab in the first line setting and found no significant difference in PFS between the arms. Both T-DM1 arms were noninferior to the taxane plus trastuzumab standard arm with a median PFS around 14 months. Although response rates were also similar between arms (around 60%), median duration of response was longer in the two T-DM1 arms (around 21 months) compared to the standard arm (12.5 months). In the T-DM1 arms, there were less at least grade 3 adverse events, fewer treatment discontinuations, and a longer maintenance of quality of life than in the standard arm [44]. Thus, T-DM1 is a better tolerated alternative to a taxane plus trastuzumab. Currently, T-DM1 is the second line standard in HER2+ ABC based on the EMILIA trial that compared T-DM1 to capecitabine and lapatinib. The final OS data from EMILIA underlined this guideline recommendation. Even though 27% of the patients in the control arm crossed over to T-DM1 upon progression, after a median follow-up of 24.1 months, median OS was still longer in the T-DM1 arm with 29.9 months compared to 25.9 months with capecitabine plus lapatinib [45]. In later treatment lines, the phase III TH3RESA trial examined the efficacy of Trastuzumab emtansine (T-DM1) versus treatment of physician's choice, that is, mostly trastuzumab plus chemotherapy, in patients with previously treated HER2+ ABC. The final OS results now demonstrate a significantly longer median OS with T-DM1 [46]. In TH3RESA, 47% of patients from the control arm crossed over to T-DM1 upon progression. Together with the EMILIA data, the OS advantages despite a substantial cross-over strongly suggest that T-DM1 is more effective in earlier treatment lines and that its impact on OS is strongest when used as early as possible and not after several anti-HER2 therapies in a late line of therapy.

The multicenter double-blind randomized HERITAGE trial ($n = 458$) evaluated a proposed trastuzumab biosimilar plus a taxane versus the originator plus the taxane in HER2+ ABC with week 24 overall response rate (ORR) being the primary endpoint. Week 24 ORR was 69.9% for the proposed biosimilar and 64.0% for the originator; ORR ratio and ORR difference were all within the predefined equivalence boundaries. Moreover, at week 48, there was also no significant efficacy difference between both treatment arms. Rates of adverse events and their distribution were also similar between both arms [47]. As further trastuzumab biosimilar trials were presented at ESMO 2017, several of these agents will now enter the approval processes in the USA and Europe. Provided that these agents fulfill all strict criteria for approval of biosimilars, they can certainly be used in clinical practice as just discussed at the ABC4 conference. They may help to increase access to anti-HER2 therapy on an international level. Nevertheless, their safety and immunogenicity will need to be monitored closely.

In summary the rate of HER2 positivity in breast cancer is about 15%. In HER2+ early breast cancer, neo-adjuvant therapy using dual antibody-based HER2 blockade together with taxane-based chemotherapy is now standard in stages II–III. In HER2+ early breast cancer, the APHINITY trial showed that dual antibody-based HER2 blockade together with adjuvant chemotherapy significantly prolongs invasive disease-free survival. And in HER2+ advanced breast cancer, T-DM1 significantly prolongs overall survival beyond first line despite substantial cross-over rates in the phase III trials.

References

1. Gnant, M., Thomssen, C., & Harbeck, N. (2015). St. Gallen/Vienna: A brief summary of the consensus discussion. *Breast Care (Basel).*, *10*, 124–130.
2. Matsen, C. B., & Neumayer, L. A. (2013). Breast cancer: A review for the general surgeon. *JAMA Surgery, 148*, 1086.
3. Reimer, T., Hartmann, S., Stachs, A., & Gerber, B. (2014). Local treatment of the axilla in early breast cancer: Concepts from the National Surgical Adjuvant Breast and Bowel Project B-04 to the Planned Intergroup Sentinel Mamma Trial. *Breast Care (Basel), 9*, 87–95.
4. Jani, A. B., Basu, A., Heimann, R., & Hellman, S. (2003). Sentinel lymph node versus axillary lymph node dissection for early-stage breast carcinoma: A comparison using a utility-adjusted number needed to treat analysis. *Cancer, 97*, 359–366.
5. Krag, D., Weaver, D., Ashikaga, T., Moffat, F., Klimberg, V. S., Shriver, C., Feldman, S., Kusminsky, R., Gadd, M., Kuhn, J., Harlow, S., & Beitsch, P. (1998). The sentinel node in breast: A multicenter validation study. *The New England Journal of Medicine, 339*, 941–946.
6. McMasters, K. M., Giuliano, A. E., Ross, M. I., Reintgen, D. S., Hunt, K. K., Byrd, D. R., Klimberg, V. S., Whitworth, P. W., Tafra, L. C., & Edwards, M. J. (1998). Sentinel-lymph-node biopsy for breast cancer-not yet the standard of care. *The New England Journal of Medicine, 339*, 990–995.
7. Reintgen, D., Giuliano, R., & Cox, C. E. (2000). Sentinel node biopsy in breast cancer: An overview. *The Breast Journal, 6*, 299–305.
8. Schwartz, G. F., Giuliano, A. E., & Veronesi, U. (2002). Consensus Conference Committee (2001) Proceedings of the consensus conference on the role of sentinel lymph node biopsy in carcinoma of the breast, April 19 to 22, Philadelphia, Pennsylvania. *Human Pathology, 33*, 579–589.
9. Veronesi, U., Paganelli, G., Viale, G., Galimberti, V., Luini, A., Zurrida, S., Robertson, C., Sacchini, V., Veronesi, P., Orvieto, E., de Cicco, C., Intra, M., Tosi, G., & Scarpa, D. (1999). Seiltinellymph node biopsy and axillary dissection in breast cancer: Results in a large series. *Journal of the National Cancer Institute, 91*, 368–373.
10. Creager, A. J., & Geisinger, K. R. (2002). Intraoperative evaluation of sentinel lymph nodes for breast carcinoma: Current methodologies. *Advances in Anatomic Pathology, 9*, 233–243.
11. Freneaux, P., Nos, C., Vincent-Salomon, A., Genin, P., Sigal-Zafrani, B., Al Ghuzian, A., Birolini, M. J., Clough, K., & Sastre-Garau, X. (2002). Histological detection of minimal metastatic involvement in axillary sentinel nodes: A rational basis for a sensitive methodology. *Modern Pathology, 15*, 641–646.
12. Silverberg, S. G. (2002). Sentinel node processing: Recommendations for pathologists. *The American Journal of Surgical Pathology, 26*, 383–385.
13. Tumer, R. R., Ollila, D. W., Stern, S., & Giuliano, A. E. (1999). Optimal histopathologic examination of the sentinel lymph node for breast carcinoma staging. *The American Journal of Surgical Pathology, 23*, 263–267.
14. Cohen, C., Alazraki, N., Styblo, T., Waldrop, S. M., Grant, S. F., & Larsen, T. (2002). Immunohistochemical evaluation of sentinel lymph nodes in breast carcinoma patients. *Applied Immunohistochemistry & Molecular Morphology, 10*, 296–303.
15. Fisher, B., Anderson, S., Bryant, J., Margolese, R. G., Deutsch, M., Fisher, E. R., et al. (2002). Twenty-year follow-up of a randomized trial comparing total mastectomy, lumpectomy, and lumpectomy plus irradiation for the treatment of invasive breast cancer. *The New England Journal of Medicine, 347*, 1233–1241.

16. Poggi, M. M., Danforth, D. N., Sciuto, L. C., Smith, S. L., Steinberg, S. M., Liewehr, D. J., et al. (2003). Eighteen-year results in the treatment of early breast carcinoma with mastectomy versus breast conservation therapy: The National Cancer Institute randomized trial. *Cancer, 98*, 697–702.
17. McGale, P., Taylor, C., Correa, C., Cutter, D., Duane, F., Ewertz, M., et al. (2014). Effect of radiotherapy after mastectomy and axillary surgery on 10-year recurrence and 20-year breast cancer mortality: Meta-analysis of individual patient data for 8135 women in 22 randomised trials. *Lancet (London England), 383*, 2127–2135.
18. Leong, B. D., Chuah, J. A., Kumar, V. M., & Yip, C. H. (2007). Breast cancer in Sabah, Malaysia: A two year prospective study. *Asian Pacific Journal of Cancer Prevention, 8*, 525–529.
19. Vorobiof, D. A., Sitas, F., & Vorobiof, G. (2001). Breast cancer incidence in South Africa. *Journal of Clinical Oncology, 19*, 125S–127S.
20. Tfayli, A., Temraz, S., Abou Mrad, R., & Shamseddine, A. (2010). Breast cancer in low- and middle-income countries: An emerging and challenging epidemic. *Journal of Oncology, 2010*, 490631.
21. Steinauer, K., Gross, M. W., Huang, D. J., Eppenberger-Castori, S., & Guth, U. (2014). Radiotherapy in patients with distant metastatic breast cancer. *Radiation Oncology (London, England), 9*, 126.
22. Foulkes, W. D., Smith, I. E., & Reis-Filho, J. S. (2010). Triple-negative breast cancer. *The New England Journal of Medicine, 363*, 1938–1948.
23. Barnes, D. M., & Hanby, A. M. (2001). Estrogens and progesterone receptors in breast cancer, past, present and future. *Histopathology, 38*, 271–274.
24. Harvey, J. M., Clark, G. M., Osborne, C. K., & Allred, D. C. (1999). Estrogen receptor status by immunohistochemistry is superior to the ligand-binding assay for predicting response to adjuvant endocrine therapy in breast cancer. *Journal of Clinical Oncology, 17*, 1474–1481.
25. Regitnig, P., Reiner, A., Dinges, H. P., Hofler, G., Muller-Holzner, E., Lax, S., Obrist, P., Rudas, M., & Quehenberger, F. (2002). Quality assurance for detection of estrogen and progesterone receptor by immunohistochemistry in Austrian pathology laboratories. *Virchows Archiv, 441*, 328–334.
26. Francis, P. A., Regan, M. M., Fleming, G. F., Láng, I., Ciruelos, E., Bellet, M., Bonnefoi, H. R., Climent, M. A., Da Prada, G. A., Burstein, H. J., Martino, S., Davidson, N. E., Geyer, C. E., Jr., Walley, B. A., Coleman, R., Kerbrat, P., Buchholz, S., Ingle, J. N., Winer, E. P., Rabaglio-Poretti, M., Maibach, R., Ruepp, B., Giobbie-Hurder, A., Price, K. N., Colleoni, M., Viale, G., Coates, A. S., Goldhirsch, A., Gelber, R. D., & SOFT Investigators, International Breast Cancer Study Group. (2015). Adjuvant ovarian suppression in premenopausal breast cancer. *The New England Journal of Medicine, 372*, 436–446.
27. Pagani, O., Regan, M. M., Walley, B. A., Fleming, G. F., Colleoni, M., Láng, I., Gomez, H. L., Tondini, C., Burstein, H. J., Perez, E. A., Ciruelos, E., Stearns, V., Bonnefoi, H. R., Martino, S., Geyer, C. E., Jr., Pinotti, G., Puglisi, F., Crivellari, D., Ruhstaller, T., Winer, E. P., Rabaglio-Poretti, M., Maibach, R., Ruepp, B., Giobbie-Hurder, A., Price, K. N., Bernhard, J., Luo, W., Ribi, K., Viale, G., Coates, A. S., Gelber, R. D., Goldhirsch, A., Francis, P. A., & TEXT and SOFT Investigators, International Breast Cancer Study Group. (2014). Adjuvant exemestane with ovarian suppression in premenopausal breast cancer. *The New England Journal of Medicine, 371*, 107–118.
28. Davies, C., Pan, H., Godwin, J., Gray, R., Arriagada, R., Raina, V., Abraham, M., Medeiros Alencar, V. H., Badran, A., Bonfill, X., Bradbury, J., Clarke, M., Collins, R., Davis, S. R., Delmestri, A., Forbes, J. F., Haddad, P., Hou, M. F., Inbar, M., Khaled, H., Kielanowska, J., Kwan, W. H., Mathew, B. S., Mittra, I., Müller, B., Nicolucci, A., Peralta, O., Pernas, F., Petruzelka, L., Pienkowski, T., Radhika, R., Rajan, B., Rubach, M. T., Tort, S., Urrútia, G., Valentini, M., Wang, Y., Peto, R., & Adjuvant Tamoxifen Longer Against Shorter (ATLAS) Collaborative Group. (2013). Long-term effects of continuing adjuvant tamoxifen to 10 years

versus stopping at 5 years after diagnosis of oestrogen receptor-positive breast cancer: ATLAS, a randomised trial. *Lancet, 381*, 805–816.
29. Gray, R. G., Rea, D., Handley, K., Bowden, S. J., Perry Earl, H. M., Poole, C. J., Bates, T., Chetiyawardana, S., Dewar, J. A., Fernando, I. N., Grieve, R., Nicoll, J., Rayter, Z., Robinson, A., Salman, A., Yarnold, J., Bathers, S., Marshall, A., Lee, M., & on behalf of the aTTom Collaborative Group. (2013). aTTom: Long-term effects of continuing adjuvant tamoxifen to 10 years versus stopping at 5 years in 6953 women with early breast cancer. *Journal of Clinical Oncology, 31*, abstr 5.
30. Ingle, J. N., Tu, D., Pater, J. L., Muss, H. B., Martino, S., Robert, N. J., Piccart, M. J., Castiglione, M., Shepherd, L. E., Pritchard, K. I., Livingston, R. B., Davidson, N. E., Norton, L., Perez, E. A., Abrams, J. S., Cameron, D. A., Palmer, M. J., & Goss, P. E. (2008). Intent-to-treat analysis of the placebo-controlled trial of letrozole for extended adjuvant therapy in early breast cancer: NCIC CTG MA.17. *Annals of Oncology, 19*, 877–882.
31. Slamon, D. J., Leyland-Jones, B., Shak, S., Fuchs, H., Paton, V., Bajamonde, A., Fleming, T., Eiermann, W., Wolter, J., Pegram, M., Baselga, J., & Norton, L. (2001). Use of chemotherapy plus a monoclonal antibody against HER2 for metastatic breast cancer that overexpresses HER2. *The New England Journal of Medicine, 344*, 783–792.
32. Romond, E. H., Perez, E. A., Bryant, J., Suman, V. J., Geyer, C. E., Jr., Davidson, N. E., Tan-Chiu, E., Martino, S., Paik, S., Kaufman, P. A., Swain, S. M., Pisansky, T. M., Fehrenbacher, L., Kutteh, L. A., Vogel, V. G., Visscher, D. W., Yothers, G., Jenkins, R. B., Brown, A. M., Dakhil, S. R., Mamounas, E. P., Lingle, W. L., Klein, P. M., Ingle, J. N., & Wolmark, N. (2005). Trastuzumab plus adjuvant chemotherapy for operable HER2-positive breast cancer. *The New England Journal of Medicine, 353*, 1673–1684.
33. Killelea, B. K., Chagpar, A. B., Horowitz, N. R., & Lannin, D. R. (2017). Characteristics and treatment of human epidermal growth factor receptor 2 positive breast cancer: 43,485 cases from the National Cancer Database treated in 2010 and 2011. *American Journal of Surgery, 213*, 426–432.
34. Curigliano, G., Burstein, H. J., Winer, E. P., et al. (2017). St. Gallen International Expert Consensus on the Primary Therapy of Early Breast Cancer 2017. De-escalating and escalating treatments for early-stage breast cancer: The St. Gallen International Expert Consensus Conference on the Primary Therapy of Early Breast Cancer 2017. *Annals of Oncology, 28*, 1700–1712.
35. Harbeck, N., & Gnant, M. (2017). Breast cancer. Comprehensive overview on state of the art management of early and advanced breast cancer. *Lancet, 389*, 1134–1150.
36. Rüschoff, J., Lebeau, A., Kreipe, H., et al.; Nicht-interventionelle Untersuchung (NIU) HER2 Study Group. (2017). Assessing HER2 testing quality in breast cancer: Variables that influence HER2 positivity rate from a large, multicenter, observational study in Germany. *Modern Pathology, 30*, 217–226.
37. Nitz, U. A., Gluz, O., Christgen, M., et al.; West-German Study Group (WSG)-ADAPT Investigators. (2017). De-escalation strategies in HER2-positive early breast cancer (EBC): Final analysis of the WSG-ADAPT HER2+/HR- phase II trial: Efficacy, safety, and predictive markers for 12 weeks of neoadjuvant dual blockade with trastuzumab and pertuzumab ± weekly paclitaxel. *Annals of Oncology, 28*, 2768–2772.
38. Harbeck, N., Gluz, O., Christgen, M., et al. (2017). De-escalation strategies in human epidermal growth factor receptor 2 (HER2)-positive early breast cancer (BC): Final analysis of the West German Study Group Adjuvant Dynamic Marker-Adjusted Personalized Therapy Trial optimizing risk assessment and therapy response prediction in early BC HER2- and hormone receptor-positive phase II randomized trial-efficacy, safety, and predictive markers for 12 weeks of neoadjuvant trastuzumab emtansine with or without endocrine therapy (ET) versus trastuzumab plus ET. *Journal of Clinical Oncology, 35*, 3046–3054.
39. Llombart-Cussac, A., Cortés, J., Paré, L., et al. (2017). HER2-enriched subtype as a predictor of pathological complete response following trastuzumab and lapatinib without chemotherapy

in early-stage HER2-positive breast cancer (PAMELA): An open-label, single-group, multicenter, phase 2 trial. *The Lancet Oncology, 18*, 545–554.

40. Nuciforo, P., Pascual, T., Cortés, J., et al. (2018). A predictive model of pathological response based on tumor cellularity and tumor-infiltrating lymphocytes (CelTIL) in HER2-positive breast cancer treated with chemo-free dual HER2 blockade. *Annals of Oncology, 29*, 170–177.
41. Loibl, S., Jackisch, C., Schneeweiss, A., et al.; investigators of the German Breast Group (GBG) and the Arbeitsgemeinschaft Gynäkologische Onkologie—Breast (AGO-B) study groups. (2017). Dual HER2-blockade with pertuzumab and trastuzumab in HER2-positive early breast cancer: A subanalysis of data from the randomized phase III GeparSepto trial. *Annals of Oncology,* 28, 497–504.
42. Pignot, F., Ajgal, Z., Xu, H., et al. (2017). Concurrent administration of anti-HER2 therapy and radiotherapy: Systematic review. *Radiotherapy and Oncology, 124*, 190–199.
43. Pituskin, E., Mackey, J. R., Koshman, S., et al. (2017). Multidisciplinary approach to novel therapies in cardio-oncology research (MANTICORE 101-Breast): A randomized trial for the prevention of trastuzumab-associated cardiotoxicity. *Journal of Clinical Oncology, 35*, 870–877.
44. Perez, E. A., Barrios, C., Eiermann, W., et al. (2017). Trastuzumab emtansine with or without pertuzumab versus trastuzumab plus taxane for human epidermal growth factor receptor 2-positive, advanced breast cancer: Primary results from the phase III MARIANNE study. *Journal of Clinical Oncology, 35*, 141–148.
45. Diéras, V., Miles, D., Verma, S., et al. (2017). Trastuzumab emtansine versus capecitabine plus lapatinib in patients with previously treated HER2-positive advanced breast cancer (EMILIA): A descriptive analysis of final overall survival results from a randomized, open-label, phase 3 trial. *The Lancet Oncology, 18*, 732–742.
46. Krop, I. E., Kim, S. B., Martin, A. G., et al. (2017). Trastuzumab emtansine versus treatment of physician's choice in patients with previously treated HER2-positive metastatic breast cancer (TH3RESA): Final overall survival results from a randomized open-label phase 3 trial. *The Lancet Oncology, 18*, 743–754.
47. Rugo, H. S., Barve, A., Waller, C. F., et al.; Heritage Study Investigators. (2017). Effect of a proposed trastuzumab biosimilar compared with trastuzumab on overall response rate in patients with ERBB2 (HER2)-positive metastatic breast cancer: A randomized clinical trial. *Journal of the American Medical Association, 317*, 37–47.

Chapter 4
The Present and Future of the Treatment of Hereditary Breast Cancer

Abstract Currently women that have a hereditary breast cancer predisposition face screening, chemoprevention, and risk-reducing surgery as options for preventing the disease. Bilateral preventive oophorectomy, combined with either MRI screening or a bilateral preventive mastectomy, can contribute considerably to a better life expectancy. Bilateral prophylactic mastectomy can reduce the risk of breast cancer by up to 95% in BRCA1/BRCA2 mutation carriers. There are two important works that must be considered seminal in the treatment of hereditary breast cancer mainly BRCA1/2 carriers. Węsierska-Gądek et al. have shown that using PARP inhibitors could be relevant in carriers of BRCA1\2 mutations based on the knowledge that BRCA1/2-mutant cells are hypersensitive to inactivation of poly(ADP-ribose) polymerase 1 (PARP-1). Another seminal paper published by Arun et al. in 2015 comparing three different PARP inhibitors such as the ABT-888 (Veliparib), BSI-201 (Iniparib), and AZD228 (Olaparib) demonstrated that they have anti-proliferative effects in breast cancer cell lines with BRCA1 or BRCA2 mutations. An interesting observation from this study was that AZD228 (Olaparib) was the most potent PARP inhibitor but only in those cells that have allelic loss.

Keywords Hereditary breast cancer · Screening · Chemoprevention · Risk-reducing surgery · Bilateral prophylactic mastectomy · BRCA1/BRCA2 mutation carriers · Poly(ADP-ribose) polymerase 1 (PARP-1) · ABT-888 (Veliparib) · BSI-201 (Iniparib) · AZD228 (Olaparib)

4.1 Introduction

Currently women that have a hereditary breast cancer predisposition face screening, chemoprevention, and risk-reducing surgery as options for preventing the disease [1–5].

Reports in the literature have shown that randomized trials failed to demonstrate a benefit for screening in premenopausal women. MRI is considered in women with dense breasts and with hereditary breast cancer syndromes to be the best approach [2]. The chemoprevention strategy is confined to estrogen receptor-positive tumors [1]. Bilateral preventive oophorectomy, combined with either MRI screening or a

J. Russo, *The Future of Prevention and Treatment of Breast Cancer*,
https://doi.org/10.1007/978-3-030-72815-1_4

bilateral preventive mastectomy, can contribute considerably to a better life expectancy [4, 5]. Bilateral prophylactic mastectomy can reduce the risk of breast cancer by up to 95% in BRCA1/BRCA2 mutation carriers. Bilateral prophylactic salpingo-oophorectomy is recommended for BRCA1/BRCA2 carriers and reducing the risk of both ovarian and breast cancer [5]. Total mastectomy is the most common treatment because of the uncertainties regarding margin status and disease extent if only reduction mammoplasty is performed in which occult breast cancer has been reported from 0.06% to 5.45% [2]. Breast-conserving treatment can be a choice for young BRCA1/2 mutation carriers despite a 40% risk of contralateral cancer within 10 years [5, 6]. The indications of preventive surgery are based on a thorough risk assessment, clinical characteristics of the different hereditary cancer susceptibility syndromes, the types of mutation, and the possibility of watchful waiting for early cancer detection [5, 6].

4.2 Approaches to the Treatment of Hereditary Breast Cancer

In the root of cancer is DNA repair mechanism in particular DNA double-strand break (DSB) repair that is essential for the survival of both normal and cancer cells. The pathways involved in DSB repair are homologous recombination and classic nonhomologous end-joining (NHEJ) pathway. BRCA1 gene encodes a protein that regulates all DSB repair pathways. DSBs also represent the most biologically significant lesions induced by ionizing radiation, and that impaired DSB repair leads to radiation sensitivity; therefore it has been expected that cancer patients with BRCA1 mutations should benefit from radiation therapy. However, the use of radiation therapy in hereditary breast cancer mainly BRCA1/2 carriers has been conflicting and inconclusive [7, 8].

The use of chemotherapy in women with stage I to stage III breast cancer who carried a BRCA1 mutation has shown in certain reports [9] that the 10-year overall survival was 78.3% (95% CI, 74.2–82.6%) and the 10-year breast cancer-specific survival was 84.2% (95% CI, 80.5–88.0%). In this study oophorectomy was associated with a significant reduction in all-cause mortality in the entire cohort (adjusted HR = 0.41; 95% CI, 0.24–0.69; $p = 0.0008$) and in breast cancer-specific mortality among ER-negative breast cancer patients (HR = 0.44; 95% CI, 0.22–0.89; $p = 0.02$).

Among women with breast cancer and a BRCA1 mutation, survival is greatly improved by oophorectomy due to the prevention of deaths from both breast and ovarian cancer [9]. The use of neo-adjuvant cisplatin in BRCA1-positive breast cancer patients has been shown to have some beneficial effect [10, 11]. PARP inhibitors are considered promising anticancer agents and have shown to be effective in clinical trials in hereditary breast cancer patients harboring mutations in BRCA1 and BRCA2 genes [11–13].

4.3 Emerging Modalities of Treatment of Hereditary Breast Cancer

There are two important works [11, 14] that must be considered seminal in the treatment of hereditary breast cancer mainly BRCA1/2 carriers. Węsierska-Gądek et al. [15] have shown that using PARP Inhibitors could be relevant in carriers of BRCA1\2 mutations based on the knowledge that BRCA1/2-mutant cells are hypersensitive to inactivation of poly(ADP-ribose) polymerase 1 (PARP-1). These authors showed that inhibition of PARP-1 by NU1025 was strongly cytotoxic for BRCA1-positive BT-20 cells, but not BRCA1-deficient SKBr-3 cells. These results raised the possibility that other PARP-1 inhibitors, particularly those tested in clinical trials, may be more efficacious against BRCA1-deficient SKBr-3 breast cancer cells than NU1025. The researchers studied the cytotoxicity of four PARP inhibitors that were under clinical evaluation (olaparib, rucaparib, iniparib, and AZD2461) and compared to that of NU1025. The sensitivity of breast cancer cells to the PARP-1 inhibition strongly varied. Remarkably, BRCA-1-deficient SKBr-3 cells were almost completely insensitive to NU1025, olaparib, and rucaparib, whereas BRCA1-expressing BT-20 cells were strongly affected by NU1025 even at low doses. In contrast, iniparib and AZD2461 were cytotoxic for both BT-20 and SKBr-3 cells. Of the four tested PARP-1 inhibitors, only AZD2461 strongly affected cell cycle progression. Interestingly, the anti-proliferative and pro-apoptotic potential of the tested PARP-1 inhibitors clearly correlated with their capacity to damage DNA. Further analyses revealed that proteomic signatures of the two studied breast cancer cell lines strongly differ, and a set of 197 proteins was differentially expressed in NU1025-treated BT-20 cancer cells. These results indicate that BT-20 cells may harbor an unknown defect in DNA repair pathway(s) rendering them sensitive to PARP-1 inhibition. They also imply that therapeutic applicability of PARP-1 inhibitors is not limited to BRCA mutation carriers but can be extended to patients harboring deficiencies in other components of the pathway(s) [14].

Another seminal paper published by Arun et al. in 2015 [11] comparing three different PARP inhibitors such as the ABT-888 (Veliparib), BSI-201 (Iniparib), and AZD228 (Olaparib) demonstrated that they have anti-proliferative effects in breast cancer cell lines with BRCA1 or BRCA2 mutations [11]. An interesting observation from this study was that AZD228 (Olaparib) was the most potent PARP inhibitor but only in those cells that have allelic loss. This was further confirmed by knocking down of BRCA1 or BRCA2 in TNBC cells with BRCA1 allelic loss by RNA interference enhancing the effect of AZD228 (Olaparib) growth inhibition and autophagy [11].

An excellent review by Schreiber et al. done in 2015 described the story of the PARP inhibitors [16]. Poly(ADP-ribosyl)ation is a post-translational modification catalyzed by poly(ADP-ribose) polymerases. PARP-1 is a molecular sensor of DNA breaks, playing a key role in the spatial and temporal organization of their repair, contributing to the maintenance of genome integrity and cell survival. PARP inhibition has also proved to be exquisitely efficient to kill tumor cells deficient in double strand break repair by homologous recombination, such as cells mutated for the breast cancer early onset genes BRCA1 or BRCA2 [15–17]. Inhibition of poly (ADP-ribose) polymerase (PARP) has shown marked benefit for breast cancer with homologous recombination deficiency, whether driven by defects in BRCA1, BRCA2, or other pathway components. Olaparib was approved by the FDA, and its use in breast cancer treatment has been a major emphasis. Combination with Platinum has shown promising benefit in metastatic BRCA mutated breast cancer or TNBC, while no such superiority was observed in the neo-adjuvant setting of TNBC [15]. Moreover, the utility of PARP inhibition in BRCA1/2 mutated breast cancer with different platinum-free interval was investigated. Immunotherapy combined with PARPi in breast cancer has obtained promising results, indicating the potential benefit of the combination therapy in patients with breast cancer [15]. Recent clinical trials in BRCA-mutant, metastatic breast cancer demonstrated improved outcomes with single-agent PARP inhibitors (olaparib and talazoparib) over chemotherapy [17]. However, resistance to PARP inhibitors remains a challenge. Primarily due to myelosuppression, the combination of PARP inhibitors with chemotherapy has been difficult [17]. In the phase III OlympiAD study [18], tumors shrank in about 60% of women with *BRCA* mutation-associated metastatic breast cancer who received the PARP inhibitor olaparib compared with 29% who received chemotherapy; the median time to disease progression was 7 months compared with 4.2 months, respectively. These findings are the first to demonstrate that a PARP inhibitor can improve progression-free survival in metastatic, hereditary breast cancer [19].

Although the data thus published indicate that PARP inhibitors are a significant advance in the treatment of hereditary breast cancer, still there is significant space for improvement. The impact of lifestyle interventions on cancer prevention and prognosis in BRCA carriers needs to also to be considered [18, 20].

4.4 BRCA 1/2 Mutations in Triple-Negative Breast Cancer

As described in Chap. 1 breast cancer is now classified by histopathological features and immunostaining for hormone receptor (ER, PgR) and HER2. Tumors lacking ER, PgR, and HER2 are classified as triple negative breast cancers (TNBCs) and comprise about 15% of breast cancers [21]. TNBCs have a poor prognosis and are sensitive to chemotherapy but quickly relapse and develop visceral metastasis [22, 23].

TNBCs share some properties with basal-like breast cancer (BLBC) characterized by overexpression of cytokeratins 5 and 8 and enrichment of genes involved in proliferation, DNA damage response, and cell cycle checkpoint. Interestingly BLBCs share many features with BRCA1-mutated tumors, and 75% of BRCA1-associated tumors are BLBCs [24]. The expression of this phenotype has created the term "BRCAness" because of its similarities with BRCA-related tumors [25–28]. The genomic overlap between TNBCs and BLBCs varies between 60% and 90% [29]. A good number of databases [30, 31] have identified seven TNBC subgroups: a basal-like 1 enriched for proliferation genes (e.g., AURKA, AURKB, CENPA,

CENPF, CCNA2, BIRC5) and DNA damage response genes (e.g., CHEK1, FANCA, RAD51, RAD21); a basal-like 2 displaying enrichment in EGF, MET, Wnt/β--catenin, and IGF1R pathways; an immunomodulatory group characterized by elevated expression of immune response (e.g., cytokine, IL-12, and IL-7 pathways); a mesenchymal and mesenchymal stem-like set that showed specific gene ontologies enriched in cell motility and cell differentiation pathways (e.g., TGF-β pathway); and luminal (androgen receptor) set featuring hormonally regulated pathway such as steroid synthesis, porphyrin metabolism, and androgen/estrogen metabolism [26].

The BRCAness phenotype is correlated with alterations in DNA repair pathways, and these tumors have loss of the PTEN locus [32, 33], TP53 mutations with deficiency in DNA repair mechanisms induced by BRCA1 loss of function [34, 35]. Tumors that are non-BRCA1/2 BLBC with a somatic homozygous loss of BARD1 a binding partner of BRCA1 [36] which interacts with TP53, RAD51, and H2AFX [36–39] also behave like the BRCA 1-mutated tumors. It has been shown [40] that many cases of BLBC have normal expression and localization of BRCA1 [40] but abnormal expression of other proteins involved in DNA repair, [30, 40, 41] emphasizing the complexity of a BRCAness phenotype.

The identification BRCAness phenotype [26–28] has acquired importance due to the response of these tumors to the treatment with poly(ADP-ribose) polymerase (PARP) 1, and two are involved in base excision repair (BER). PARP detects single-strand breaks, binds to the sites of DNA damage and begins its auto poly(ADP-ribosyl)ation. Such an event is responsible for the recruitment of BER proteins, such as DNA-polymerase β, DNA ligase III, and the scaffold protein XRCC1. Moreover, poly(ADP-ribosyl)ation regards other proteins, such as histones H1 and H2B, leading to chromatin condensation and the binding of DNA repair enzymes [42]. Therefore, PARP inhibition in BRCA1/2-associated breast cancer is based on the concept of "synthetic lethality" developed in *Drosophila* [43] and applied it to cancer [44]. Several clinical trials evaluating PARPi treatment in TNBCs are listed in www.clinicaltrials.gov.

4.5 Looking for New Susceptibility Genes and Possible Target of Therapy

Breast cancer is a large collection of distinct biological entities that has been clearly established by the different and variegated types of genes affected. It is in breast cancer where the concept of personalized medicine is relevant. Personal medicine is aiming to use genetic profiles either from patients or from the primary tumors to tailor therapeutic decisions [44]. The use of next-generation sequencing (NGS) in clinical diagnostics has identified large number of new germline mutations, and in a study of 255 women, 57 of them had a pathogenic mutation in BRCA1/2 genes, and 17 carried pathogenic mutations in other genes, such as PALB2, ATM, BRIP1, RAD51D, MSH6, PPM1D, RECQL4, ERCC3, TSC2, SLX4, and other Fanconi anemia genes. The important message of this study is that these patients had a much higher percentage of bilateral breast cancer (BBC) and a lower rate of ovarian cancer than BRCA-mutated patients and patients with no pathogenic mutations [45].

A large study designed to identify novel risk loci and likely causal genes was conducted in 122,977 cases at high risk and compared with 105,974 controls of European ancestry [46]. In this study the transcriptome-wide association of genetically predicted gene expression with breast cancer risk was done using data from the Genotype-Tissue Expression Project to establish genetic models to predict gene expression in breast tissue and evaluated model performance using data from The Cancer Genome Atlas. It is important to indicate that in this study, 14 genes at loci not yet reported for breast cancer were identified. The clinical relevance of this findings is still needed but clearly indicates the need for further research in this area.

In another study a genome-wide association of breast cancer in 122,977 cases and 105,974 controls of European ancestry and 14,068 cases and 13,104 controls of East Asian ancestry [47] identified 65 new loci that are associated with overall breast cancer risk at $p < 5 \times 10^{-8}$ [47].

4.6 Prevention vs Treatment of Hereditary Breast Cancer

Data published indicated that 5–10% of breast cancers are considered hereditary with an early age of onset and affect the ovaries and both breasts [45]. Among the preventive strategies have been breast self-test training and monthly breast self-examination. Large-scale clinical trials have shown the important role of chemoprevention in reducing the occurrence of hereditary breast cancer. Prophylactic mastectomy can reduce the incidence rate of breast cancer in high-risk women by 90% and decrease the breast cancer mortality rate in medium-risk and high-risk women by 100% and 81%, respectively [48]. A study reporting a web-based decision aid, BRECONDA (Breast Reconstruction Decision Aid), to assist women in decision-making regarding breast reconstruction, has shown that intervention participants reported lower decisional conflict ($p = 0.027$), and greater knowledge ($p = 0.019$) and satisfaction with information ($p < 0.0005$) at 2-months follow-up compared with controls. Intervention participants reported high user acceptability and satisfaction with the intervention [49]. Genetic counseling is playing an important role in the management of women with familial breast cancer mainly those carriers of BRCA1/2 mutations. It seems that the main concern is the cost of the testing [50, 51].

References Cited

1. Jatoi, I., & Benson, J. R. (2016). Management of women with a hereditary predisposition for breast cancer. *Future Oncology, 12*, 2277–2288.
2. Carlson, G. W. (2016). The management of breast cancer detected by reduction mammaplasty. *Clinics in Plastic Surgery, 43*, 341–347.

3. Bit-Sava, E. M. (2014). Genetical aspects of surgical treatment in patients with hereditary breast cancer. *Vestnik Khirurgii Imeni I. I. Grekova, 173*, 27–30.
4. Saadatmand, S., Obdeijn, I. M., Koppert, L. B., & Tilanus-Linthorst, M. (2013). Choices for women at risk of hereditary breast cancer. *Nederlands Tijdschrift voor Geneeskunde, 157*, A6625.
5. Noguès, C., & Mouret-Fourme, E. (2012). Prophylactic surgery in common hereditary cancer syndromes. *Bulletin de l'Académie Nationale de Médecine, 196*, 1237–1245.
6. Lynch, H. T., Snyder, C., & Lynch, J. (2012). Hereditary breast cancer: practical pursuit for clinical translation. *Annals of Surgical Oncology, 19*, 1723–1731.
7. Cooper, B. T., Murphy, J. O., Sacchini, V., & Formenti, S. C. (2013). Local approaches to hereditary breast cancer. *Annals of Oncology, 24*(Suppl 8), viii54–viii60.
8. Kan, C., & Zhang, J. (2015). BRCA1 mutation: a predictive marker for radiation therapy? *International Journal of Radiation Oncology, Biology, Physics, 93*, 281–293.
9. Huzarski, T., Byrski, T., Gronwald, J., Cybulski, C., Oszurek, O., Szwiec, M., Gugała, K., Stawicka, M., Morawiec, Z., Mierzwa, T., Falco, M., Janiszewska, H., Kilar, E., Marczyk, E., Kozak-Klonowska, B., Siołek, M., Surdyka, D., Wiśniowski, R., Posmyk, M., Domagała, P., Sun, P., Lubiński, J., Narod, S. A., & Polish Breast Cancer Consortium. (2016). The impact of ophorectomy on survival after breast cancer in BRCA1-positive breast cancer patients. *Breast Cancer Research and Treatment, 156*, 371–378.
10. Byrski, T., Huzarski, T., Dent, R., Marczyk, E., Jasiowka, M., Gronwald, J., Jakubowicz, J., Cybulski, C., Wisniowski, R., Godlewski, D., Lubinski, J., & Narod, S. A. (2014). Pathologic complete response to neoadjuvant cisplatin in BRCA1-positive breast cancer patients. *Breast Cancer Research and Treatment, 147*, 401–405.
11. Arun, B., Akar, U., Gutierrez-Barrera, A. M., Hortobagyi, G. N., & Ozpolat, B. (2015). The PARP inhibitor AZD2281 (Olaparib) induces autophagy/mitophagy in BRCA1 and BRCA2 mutant breast cancer cells. *International Journal of Oncology, 47*, 262–268.
12. Drost, R., & Jonkers, J. (2014). Opportunities and hurdles in the treatment of BRCA1-related breast cancer. *Oncogene, 33*, 3753–3763.
13. Domchek, S. M., Mitchell, G., Lindeman, G. J., Tung, N. M., Balmaña, J., Isakoff, S. J., Schmutzler, R., Audeh, M. W., Loman, N., Scott, C., Friedlander, M., Kaufman, B., Garber, J. E., Tutt, A., & Robson, M. E. (2011). Challenges to the development of new agents for molecularly defined patient subsets: lessons from BRCA1/2-associated breast cancer. *Journal of Clinical Oncology, 29*, 4224–4226.
14. Węsierska-Gądek, J., Mauritz, M., Mitulovic, G., & Cupo, M. (2015). Differential potential of pharmacological PARP inhibitors for inhibiting cell proliferation and inducing apoptosis in human breast cancer cells. *Journal of Cellular Biochemistry, 116*, 2824–2839.
15. Wang, X., Shi, Y., Huang, D., & Guan, X. (2018). Emerging therapeutic modalities of PARP inhibitors in breast cancer. *Cancer Treatment Reviews, 68*, 62–68.
16. Schreiber, V., Illuzzi, G., Héberlé, E., & Dantzer, F. (2015). From poly(ADP-ribose) discovery to PARP inhibitors in cancer therapy. *Bulletin du Cancer, 102*, 863–873.
17. Turk, A. A., & Wisinski, K. B. (2018). PARP inhibitors in breast cancer: Bringing synthetic lethality to the bedside. *Cancer, 124*, 2498–2506.
18. Bayraktar, S., & Glück, S. (2012). Systemic therapy options in BRCA mutation-associated breast cancer. *Breast Cancer Research and Treatment, 135*, 355–366.
19. Olaparib keeps hereditary breast tumors in check. (2017). *Cancer Discover, 7*, 10.
20. Kiechle, M., Dukatz, R., Yahiaoui-Doktor, M., Berling, A., Basrai, M., Staiger, V., Niederberger, U., Marter, N., Lammert, J., Grill, S., Pfeifer, K., Rhiem, K., Schmutzler, R. K., Laudes, M., Siniatchkin, M., Halle, M., Bischoff, S. C., & Engel, C. (2017). Feasibility of structured endurance training and Mediterranean diet in BRCA1 and BRCA2 mutation carriers – an interventional randomized controlled multicenter trial (LIBRE-1). *BMC Cancer, 17*, 752.
21. Rakha, E. A., & Chan, S. (2011). Metastatic triple-negative breast cancer. *Clinical Oncology (Royal College of Radiologists), 23*, 587–600.
22. Haffty, B. G., Yang, Q., Reiss, M., et al. (2006). Locoregional relapse and distant metastasis in conservatively managed triple negative early-stage breast cancer. *Journal of Clinical Oncology, 24*, 5652–5657.

23. Dent, R., Trudeau, M., Pritchard, K. I., et al. (2007). Triple-negative breast cancer: Clinical features and patterns of recurrence. *Clinical Cancer Research, 13*, 4429–4434.
24. Waddell, N., Arnold, J., Cocciardi, S., et al. (2010). Subtypes of familial breast tumors revealed by expression and copy number profiling. *Breast Cancer Research and Treatment, 123*, 661–677.
25. Kreike, B., van Kouwenhove, M., Horlings, H., et al. (2007). Gene expression profiling and histopathological characterization of triple-negative/basal-like breast carcinomas. *Breast Cancer Research, 9*, R65.
26. De Summa, S., Pinto, R., Sambiasi, D., Petriella, D., Paradiso, V., Paradiso, A., & Tommasi, S. (2013). BRCAness: a deeper insight into basal-like breast tumors. *Annals of Oncology, 24*(Suppl 8), viii13–viii21.
27. Afghahi, A., Telli, M. L., & Kurian, A. W. (2016). Genetics of triple-negative breast cancer: Implications for patient care. *Current Problems in Cancer, 40*, 130–140.
28. Domagala, P., Hybiak, J., Cybulski, C., & Lubinski, J. (2017). BRCA1/2-negative hereditary triple-negative breast cancers exhibit BRCAness. *International Journal of Cancer, 140*, 1545–1550.
29. Al Tamimi, D. M., Shawarby, M. A., Ahmed, A., et al. (2010). Protein expression profile and prevalence pattern of the molecular classes of breast cancer—A Saudi population based study. *BMC Cancer, 10*, 223.
30. Lehmann, B. D., Bauer, J. A., Chen, X., et al. (2011). Identification of human triple-negative breast cancer subtypes and preclinical models for selection of targeted therapies. *The Journal of Clinical Investigation, 121*, 2750–2767.
31. Bertucci, F., Finetti, P., Cervera, N., et al. (2008). How basal are triple-negative breast cancers? *International Journal of Cancer, 123*, 236–240.
32. Saal, L. H., Gruvberger-Saal, S. K., Persson, C., et al. (2008). Recurrent gross mutations of the PTEN tumor suppressor gene in breast cancers with deficient DSB repair. *Nature Genetics, 40*, 102–107.
33. Mendes-Pereira, A. M., Martin, S. A., Brough, R., et al. (2009). Synthetic lethal targeting of PTEN mutant cells with PARP inhibitors. *EMBO Molecular Medicine, 1*, 315–322.
34. Kusinska, R., Potemski, P., Jesionek-Kupnicka, D., et al. (2005). Immunohistochemical identification of basal-type cytokeratins in invasive ductal breast carcinoma—Relation with grade, stage, estrogen receptor and HER2. *Polish Journal of Pathology, 56*, 107–110.
35. Holstege, H., Joosse, S. A., van Oostrom, C. T., et al. (2009). High incidence of protein truncating TP53 mutations in BRCA1-related breast cancer. *Cancer Research, 69*, 3625–3633.
36. Mallery, D. L., Vandenberg, C. J., & Hiom, K. (2002). Activation of the E3 ligase function of the BRCA1/BARD1 complex by polyubiquitin chains. *The EMBO Journal, 21*, 6755–6762.
37. Wu, L. C., Wang, Z. W., Tsan, J. T., et al. (1996). Identification of a RING protein that can interact in vivo with the BRCA1 gene product. *Nature Genetics, 14*, 430–440.
38. Irminger-Finger, I., Soriano, J. V., Vaudan, G., et al. (1998). In vitro repression of Brca1-associated RING domain gene, Bard1, induces phenotypic changes in mammary epithelial cells. *The Journal of Cell Biology, 143*, 1329–1339.
39. Chen, A., Kleiman, F. E., Manley, J. L., et al. (2002). Autoubiquitination of the BRCA1*BARD1 RING ubiquitin ligase. *The Journal of Biological Chemistry, 277*, 22085–22092.
40. Wang, W. (2007). Emergence of a DNA-damage response network consisting of Fanconi anaemia and BRCA proteins. *Nature Reviews. Genetics, 8*, 735–748.
41. Alli, E., Sharma, V. B., Sunderesakumar, P., et al. (2009). Defective repair of oxidative DNA damage in triple-negative breast cancer confers sensitivity to inhibition of poly (ADP-ribose) polymerase. *Cancer Research, 69*, 3589–3596.
42. de Murcia, G., Huletsky, A., Lamarre, D., et al. (1986). Modulation of chromatin superstructure induced by poly(ADP-ribose) synthesis and degradation. *The Journal of Biological Chemistry, 261*, 7011–7017.

43. Dobzhansky, T. (1946). Genetics of natural populations. Xiii. Recombination and variability in populations of Drosophila Pseudoobscura. *Genetics, 31*, 269–290.
44. Gonçalves, A., Moretta, J., & Eisinger, F. (2013). Bertucci F[Personalized medicine and breast cancer: anticipatory medicine, prognostic evaluation and therapeutic targeting]. *Bulletin du Cancer, 100*, 1295–1310.
45. Tedaldi, G., Tebaldi, M., Zampiga, V., Danesi, R., Arcangeli, V., Ravegnani, M., Cangini, I., Pirini, F., Petracci, E., Rocca, A., Falcini, F., Amadori, D., & Calistri, D. (2017). Multiple-gene panel analysis in a case series of 255 women with hereditary breast and ovarian cancer. *Oncotarget, 8*, 47064–47075.
46. Wu L, Shi W, Long J, Guo X, Michailidou K, Beesley J, Bolla MK, Shu XO, Lu Y, Cai Q, Al-Ejeh F, Rozali E, Wang Q, Dennis J, Li B, Zeng C, Feng H, Gusev A, Barfield RT, Andrulis IL, Anton-Culver H, Arndt V, Aronson KJ, Auer PL, Barrdahl M, Baynes C, Beckmann MW, Benitez J, Bermisheva M, Blomqvist C, Bogdanova NV, Bojesen SE, Brauch H, Brenner H, Brinton L, Broberg P, Brucker SY, Burwinkel B, Caldés T, Canzian F, Carter BD, Castelao JE, Chang-Claude J, Chen X, Cheng TD, Christiansen H, Clarke CL, NBCS Collaborators, Collée M, Cornelissen S, Couch FJ, Cox D, Cox A, Cross SS, Cunningham JM, Czene K, Daly MB, Devilee P, Doheny KF, Dörk T, Dos-Santos-Silva I, Dumont M, Dwek M, Eccles DM, Eilber U, Eliassen AH, Engel C, Eriksson M, Fachal L, Fasching PA, Figueroa J, Flesch-Janys D, Fletcher O, Flyger H, Fritschi L, Gabrielson M, Gago-Dominguez M, Gapstur SM, García-Closas M, Gaudet MM· Ghoussaini M, Giles GG, Goldberg MS, Goldgar DE, González-Neira A, Guénel P, Hahnen E, Haiman CA, Håkansson N, Hall P, Hallberg E, Hamann U, Harrington P, Hein A, Hicks B, Hillemanns P, Hollestelle A, Hoover RN, Hopper JL, Huang G, Humphreys K, Hunter DJ, Jakubowska A, Janni W, John EM, Johnson N, Jones K, Jones ME, Jung A, Kaaks R, Kerin MJ, Khusnutdinova E, Kosma VM, Kristensen VN, Lambrechts D, Le Marchand L, Li J, Lindström S, Lissowska J, Lo WY, Loibl S, Lubinski J, Luccarini C, Lux MP, MacInnis RJ, Maishman T, Kostovska IM, Mannermaa A, Manson JE, Margolin S, Mavroudis D, Meijers-Heijboer H, Meindl A, Menon U, Meyer J, Mulligan AM, Neuhausen SL, Nevanlinna H, Neven P, Nielsen SF, Nordestgaard BG, Olopade OI, Olson JE, Olsson H, Peterlongo P, Peto J, Plaseska-Karanfilska D, Prentice R, Presneau N, Pylkäs K, Rack B, Radice P, Rahman N, Rennert G, Rennert HS, Rhenius V, Romero A, Romm J, Rudolph A, Saloustros E, Sandler DP, Sawyer EJ, Schmidt MK, Schmutzler RK, Schneeweiss A, Scott RJ, Scott CG, Seal S, Shah M, Shrubsole MJ, Smeets A, Southey MC, Spinelli JJ, Stone J, Surowy H, Swerdlow AJ, Tamimi RM, Tapper W, Taylor JA, Terry MB, Tessier DC, Thomas A, Thöne K, Tollenaar RAEM, Torres D, Truong T, Untch M, Vachon C, Van Den Berg D, Vincent D, Waisfisz Q, Weinberg CR, Wendt C, Whittemore AS, Wildiers H, Willett WC, Winqvist R, Wolk A, Xia L, Yang XR, Ziogas A, Ziv E; kConFab/AOCS Investigators, Dunning AM, Pharoah PDP, Simard J, Milne RL, Edwards SL, Kraft P, Easton DF, Chenevix-Trench G, Zheng W. (2018). A transcriptome-wide association study of 229,000 women identifies new candidate susceptibility genes for breast cancer. *Nature Genetics, 50*, 968–978.
47. Harter P, Hauke J, Heitz F, Reuss A, Kommoss S, Marmé F, Heimbach A, -Michailidou K, Lindström S, Dennis J, Beesley J, Hui S, Kar S, Lemaçon A, Soucy P, Glubb D, Rostamianfar A, Bolla MK, Wang Q, Tyrer J, Dicks E, Lee A, Wang Z, Allen J, Keeman R, Eilber U, French JD, Qing Chen X, Fachal L, McCue K, McCart Reed AE, Ghoussaini M, Carroll JS, Jiang X, Finucane H, Adams M, Adank MA, Ahsan H, Aittomäki K, Anton-Culver H, Antonenkova NN, Arndt V, Aronson KJ, Arun B, Auer PL, Bacot F, Barrdahl M, Baynes C, Beckmann MW, Behrens S, Benitez J, Bermisheva M, Bernstein L, Blomqvist C, Bogdanova NV, Bojesen SE, Bonanni B, Børresen-Dale AL, Brand JS, Brauch H, Brennan P, Brenner H, Brinton L, Broberg P, Brock IW Broeks A, Brooks-Wilson A, Brucker SY, Brüning T, Burwinkel B, Butterbach K, Cai Q, Cai H, Caldés T, Canzian F, Carracedo A, Carter BD, Castelao JE, Chan TL, David Cheng TY, Seng Chia K, Choi JY, Christiansen H, Clarke CL; NBCS Collaborators, Collée M, Conroy DM, Cordina-Duverger E, Cornelissen S, Cox DG, Cox A, Cross SS, Cunningham JM, Czene K, Daly MB, Devilee P, Doheny KF, Dörk T, Dos-Santos-Silva I, Dumont M, Durcan L, Dwek M, Eccles DM, Ekici AB, Eliassen AH, Ellberg C, Elvira M,

Engel C, Eriksson M, Fasching PA, Figueroa J, Flesch-Janys D, Fletcher O, Flyger H, Fritschi L, Gaborieau V, Gabrielson M, Gago-Dominguez M, Gao YT, Gapstur SM, García-Sáenz JA, Gaudet MM, Georgoulias V, Giles GG, Glendon G, Goldberg MS, Goldgar DE, González-Neira A, Grenaker Alnæs GI, Grip M, Gronwald J, Grundy A, Guénel P, Haeberle L, Hahnen E, Haiman CA, Håkansson N, Hamann U, Hamel N, Hankinson S, Harrington P, Hart SN, Hartikainen JM, Hartman M, Hein A, Heyworth J, Hicks B, Hillemanns P, Ho DN, Hollestelle A, Hooning MJ, Hoover RN, Hopper JL, Hou MF, Hsiung CN, Huang G, Humphreys K, Ishiguro J, Ito H, Iwasaki M, Iwata H, Jakubowska AJanni W, John EM, Johnson N, Jones K, Jones M, Jukkola-Vuorinen A, Kaaks R, Kabisch M, Kaczmarek K, Kang D, Kasuga Y, Kerin MJ, Khan S, Khusnutdinova E, Kiiski JI, Kim SW, Knight JA, Kosma VM, Kristensen VN, Krüger U, Kwong A, Lambrechts D, Le Marchand L, Lee E, Lee MH, Lee JW, Neng Lee C, Lejbkowicz F, Li J, Lilyquist J, Lindblom A, Lissowska J, Lo WY, Loibl S, Long J, Lophatananon A, Lubinski J, Luccarini C, Lux MP, Ma ESK, MacInnis RJ, Maishman T, Makalic E, Malone KE, Kostovska IM, Mannermaa A, Manoukian S, Manson JE, Margolin S, Mariapun S, Martinez ME, Matsuo K, Mavroudis D, McKay J, McLean C, Meijers-Heijboer H, Meindl A, Menéndez P, Menon U, Meyer J, Miao H, Miller N, Taib NAM, Muir K, Mulligan AM, Mulot C, Neuhausen SL, Nevanlinna H, Neven P, Nielsen SF, Noh DY, Nordestgaard BG, Norman A, Olopade OI, Olson JE, Olsson H, Olswold C, Orr N, Pankratz VS, Park SK, Park-Simon TW, Lloyd R, Perez JIA, Peterlongo P, Peto J, Phillips KA, Pinchev M, Plaseska-Karanfilska D, Prentice R, Presneau N, Prokofyeva D, Pugh E, Pylkäs K, Rack B, Radice P, Rahman N, Rennert G, Rennert HS, Rhenius V, Romero A, Romm J, Ruddy KJ, Rüdiger T, Rudolph A, Ruebner M, Rutgers EJT, Saloustros E, Sandler DP, Sangrajrang S, Sawyer EJ, Schmidt DF, Schmutzler RK, Schneeweiss A, Schoemaker MJ, Schumacher F, Schürmann P, Scott RJ, Scott C, Seal S, Seynaeve C, Shah M, Sharma P, Shen CY, Sheng G, Sherman ME, Shrubsole MJ, Shu XO, Smeets A, Sohn C, Southey MC, Spinelli JJ, Stegmaier C, Stewart-Brown S, Stone J, Stram DO, Surowy H, Swerdlow A, Tamimi R, Taylor JA, Tengström M, Teo SH, Beth Terry M, Tessier DC, Thanasitthichai S, Thöne K, Tollenaar RAEM, Tomlinson I, Tong L, Torres D, Truong T, Tseng CC, Tsugane S, Ulmer HU, Ursin G, Untch M, Vachon C, van Asperen CJ, Van Den Berg D, van den Ouweland AMW, van der Kolk L, van der Luijt RB, Vincent D, Vollenweider J, Waisfisz Q, Wang-Gohrke S, Weinberg CR, Wendt C, Whittemore AS, Wildiers H, Willett W, Winqvist R, Wolk A, Wu AH, Xia L, Yamaji T, Yang XR, Har Yip C, Yoo KY, Yu JC, Zheng W, Zheng Y, Zhu B, Ziogas A, Ziv E; ABCTB Investigators; ConFab/AOCS Investigators, Lakhani SR, Antoniou AC, Droit A, Andrulis IL, Amos CI, Couch FJ, Pharoah PDP, Chang-Claude J, Hall P, Hunter DJ, Milne RL, García-Closas M, Schmidt MK, Chanock SJ, Dunning AM, Edwards SL, Bader GD, Chenevix-Trench G, Simard J, Kraft P, Easton DF. (2017). Association analysis identifies 65 new breast cancer risk loci. Nature, *551*, 92–94.

48. Cao, A., Huang, L., & Shao, Z. (2017). The preventive intervention of hereditary breast cancer. *Advances in Experimental Medicine and Biology, 1026*, 41–57.
49. Sherman, K. A., Kilby, C. J., Shaw, L. K., Winch, C., Kirk, J., Tucker, K., & Elder, E. (2017). Facilitating decision-making in women undergoing genetic testing for hereditary breast cancer: BRECONDA randomized controlled trial results. *Breast, 36*, 79–85.
50. Steffen, L. E., Du, R., Gammon, A., Mandelblatt, J. S., Kohlmann, W. K., Lee, J. H., Buys, S. S., Stroup, A. M., Campo, R. A., Flores, K. G., Vicuña, B., Schwartz, M. D., & Kinney, A. Y. (2017). Genetic testing in a population-based sample of breast and ovarian cancer survivors from the REACH randomized trial: cost barriers and moderators of counseling mode. *Cancer Epidemiology, Biomarkers & Prevention, 26*, 1772–1780.
51. Kwiatkowski, F., Dessenne, P., Laquet, C., Daures, J. P., Gay-Bellile, M., & Bignon, Y. J. (2016). BRACAVENIR – impact of a psychoeducational intervention on expectations and coping in young women (aged 18-30 years) exposed to a high familial breast/ovarian cancer risk: study protocol for a randomized controlled trial. *Trials, 17*, 509.

Chapter 5
Basis for the Epigenetic Treatment of Triple-Negative Breast Cancer

Abstract Triple-negative breast cancer (TNBC) is an aggressive neoplasia with no effective therapy. Our laboratory has developed a unique TNBC cell model presenting epithelial mesenchymal transition (EMT), a process known to be important for tumor progression and metastasis. There is increasing evidence showing that epigenetic mechanisms are involved in the activation of EMT. The objective of this study is to epigenetically reverse the process of EMT in TNBC by using DNA methyltransferase inhibitors (DNMTi) and histone deacetylase inhibitors (HDACi). We evaluated the antitumor effect of three DNMTi and six HDACi using our TNBC cell model by MTT assay, migration and invasion assay, three-dimensional culture, and colony formation assay. We then performed the combined treatment both in vitro and in vivo using the most potent DNMTi and HDACi and tested the combined treatment in a panel of breast cancer cell lines. We investigated changes of EMT markers and potential signaling pathways associated with the antitumor effects. We showed that DNMTi and HDACi can reprogram highly aggressive TNBC cells that have undergone EMT to a less aggressive phenotype. SGI-110 and MS275 are superior to other seven compounds being tested. The combination of SGI with MS275 exerts a greater effect than single agent alone in inhibiting cell proliferation, motility, colony formation, and stemness of cancer cells. We also demonstrated that MS275 and the combination of SGI with MS275 exert in vivo antitumor effect. We revealed that the combined treatment synergistically reverses EMT through inhibiting EpCAM cleavage and WNT signaling, suppressing mutant p53, ZEB1, and EZH2, and inducing E-cadherin, apoptosis, as well as histone H3 tri-methylation. Our study showed that DNMTi and HDACi exert antitumor activity in TNBC cells partially by epigenetically reprograming EMT. Our findings strongly suggest that TNBC is sensitive to epigenetic therapies. Therefore, we propose a new strategy to treat TNBC by using the combination of SGI-110 with MS275, which exerts superior antitumor effects by simultaneously targeting multiple pathways.

Keywords DNA methyltransferase · Epithelial mesenchymal transition · Histone deacetylase · Triple negative breast cancer

J. Russo, *The Future of Prevention and Treatment of Breast Cancer*,
https://doi.org/10.1007/978-3-030-72815-1_5

5.1 Introduction to the Problem

Triple-negative breast cancer (TNBC) is a subtype of breast cancer negative for hormone receptors and human epidermal growth factor receptor 2 and compared to other breast cancer subtypes, is more aggressive and more difficult to treat. Radiotherapy and chemotherapy are the two main therapies combined with surgery for TNBC [1]. Standard chemotherapies show some effects for subtypes of TNBC; however, patients with advanced disease typically respond poorly to current chemotherapies [2]. Though some potential therapies are in clinical trial [3, 4], there is still no effective treatment available.

A crucial step in cancer progression is epithelial-mesenchymal transition (EMT) that precedes tumor metastasis [5, 6] and therefore creating an important biological target for therapy. Studies in vitro have shown that EMT can be induced in breast cancer cell lines by the treatments including TGF beta [7, 8] and WNT [9]. In addition, the EMT is associated with activation of stem cells containing the $CD44^{+}CD24^{-/low}$ phenotype [10–12] and are significantly higher in basal-like and mesenchymal-like subtype associated with poor outcome [13–15]. Although EMT is a highly molecular complex with a cascade of gene activation [16–20], there is increasing evidence showing that epigenetic mechanisms are involved in the activation of this process. The best example is the promoter hyper-methylation of E-cadherin *(CDH1) gene,* resulting in breast tumor cells with a more aggressive phenotype [21]. E-cadherin expression is epigenetically regulated in the 5' CpG island of *CDH1* gene and is downregulated in breast cancer cell lines and primary breast cancer tissues [22]. The importance of this finding is that E-cadherin expression is also repressed by a number of EMT inducers including SNAIL, SLUG, ZEB1, ZEB2, and TWIST [16–19]. For example, SNAIL interacts with *CDH1* promoter and recruits histone deacetylase (HDAC) containing complexes to *CDH1* promoter, leading to deacetylation of histone H3/H4 and inducing transcriptional repression of E-cadherin [16]. Also SLUG expression is correlated with the absence of E-cadherin [17] transcripts and associated with HDAC [23]. ZEB1/ZEB2 repress E-cadherin transcription by recruiting the CtBP corepressor complex that contains ZEB1/ZEB2, HDAC1/2, the histone methyltransferase G9a, and some other proteins [24, 25]. TWIST also interacts with multiple components including HDAC2 to repress E-cadherin, inducing EMT and cell invasion and metastasis [26].

The knowledge that EMT is epigenetically regulated is of great therapeutic relevance because it indicates that EMT could be reversible. Our studies performed in the Irma H. Russo, MD, Breast Cancer Research Laboratory at the FCCC in Philadelphia have shown that DNA methyltransferase inhibitor (DNMTi) and histone deacetylase inhibitor (HDACi) have antitumor effect on TNBC cells by reprograming EMT.

5.2 The Model of Triple-Negative Breast Cancer

There are numerous cell lines that have been obtained from TNBC; however there is a lack of reliable models that can trace the early events in the transformation of a normal basal cell to the highly metastatic type that is characteristic of this type of cancer. In Fig. 5.1 the location of the basal cell in the ductular structure of the human breast is shown.

We have further characterized that MCF10F and MCF10A cells [27] are basal cells (Fig. 5.2), and established a model of TNBC cells [20] which consists of the normal like human breast epithelial cell line MCF10F, the cell line trMCF which was transformed from MCF10F, and the tumorigenic cell line bsMCF derived from trMCF, as well as two highly tumorigenic and metastatic cancer cell lines XtMCF and LmMCF developed from bsMCF [12] (Fig. 5.3).

The bsMCF, XtMCF, and LmMCF cells have undergone EMT, displaying mesenchymal phenotype [12]. The advantage of this unique cell model is that all the cells are derived from the same genotype present in the MCF10F cells and we have been able to identify epigenetic and genomic changes during the process of neoplastic transformation. In Fig. 5.4 is depicted the number of hyper- and hypo-methylated genes during the process of EMT. In Fig. 5.5 is shown the methylation status of three important genes involved in the EMT, EPCAM, CDH1, and vimentin.

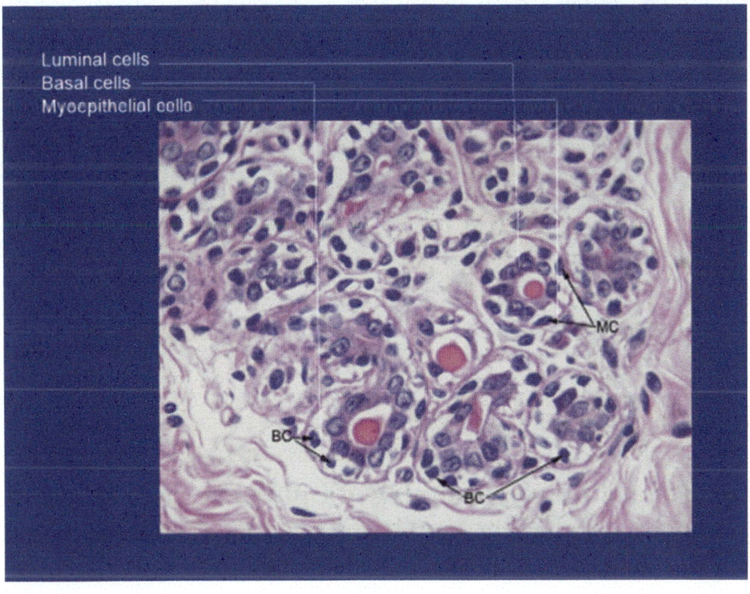

Fig. 5.1 Histological section of the ductular structures of the human breast showing the three main cell types, the luminal, the basal and the myoepithelial cells. Paraffin section stained with hematoxylin and eosin, 40×. (From Russo and Russo [71])

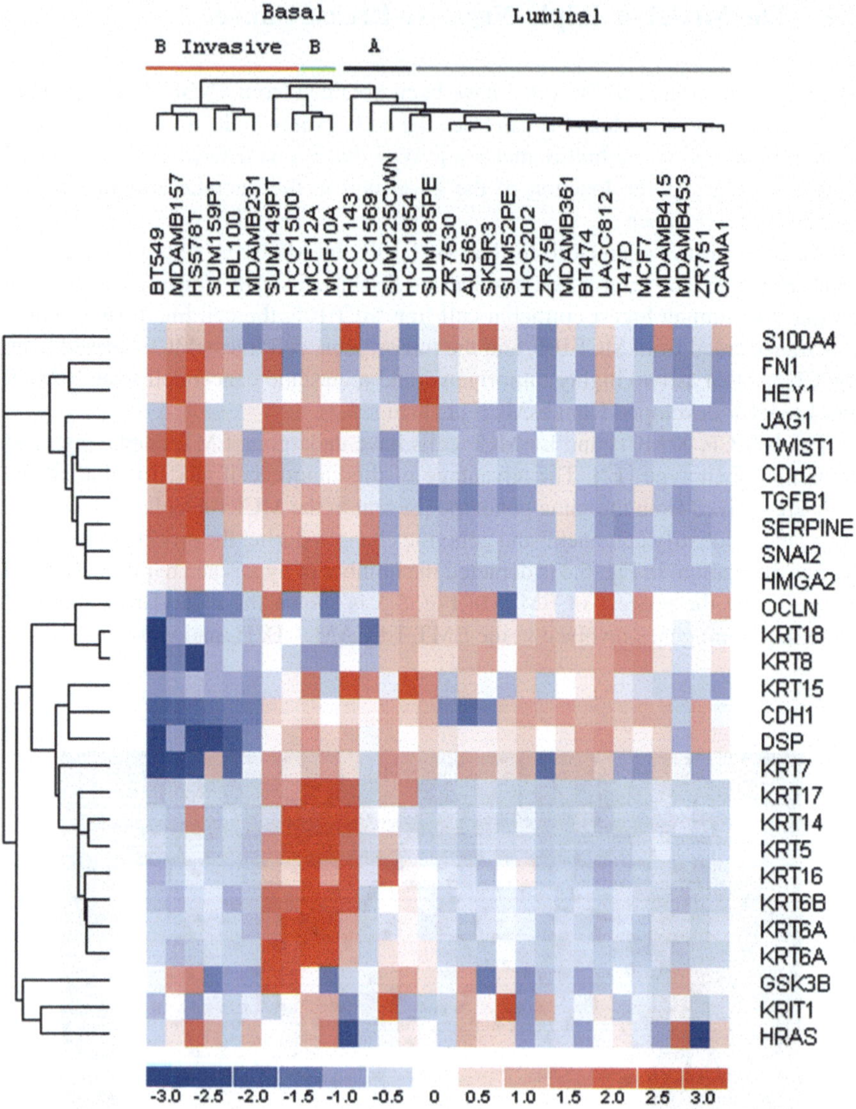

Fig. 5.2 Heatmap showing the high expression profile of the MCF10A and MCF10F cells containing the markers identified in the basal cell type

5.3 Epigenetic Reprogramming of Epithelial Mesenchymal Transition in Triple-Negative Breast Cancer Cells

Using this unique experimental model [12] leads us to explore the potential use of demethylating agents as therapeutic drugs against triple-negative breast cancer. We first evaluated the antitumor efficacy of a panel of DNMTi and HDACi by different

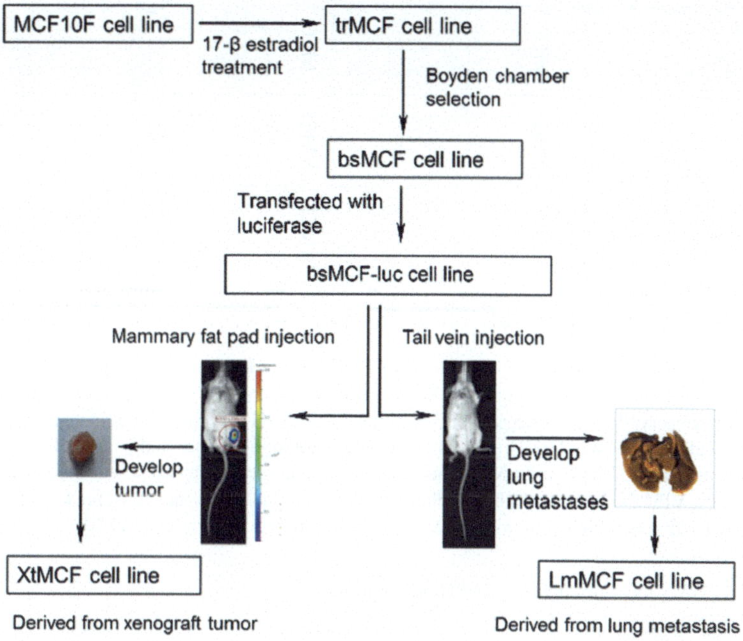

Fig. 5.3 Model of triple-negative breast cancer. (From Su et al. [12])

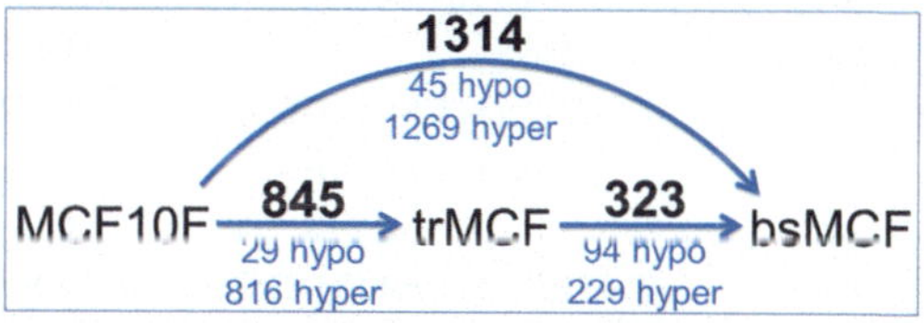

Fig. 5.4 The methylation status of the genes in breast epithelial cells during the process of EMT was detected by using magnetic beads coated in MBD (methyl binding domain); polypeptide binds to methylated DNA. In this way the CpG-methylated DNA is captured by the MBD protein, and the methylated DNA is eluted and sequenced. Number of differentially methylated regions found in each comparison is shown in the graph. (From Santucci-Pereira et al. [72])

assays, and then performed the combined treatment using the selected compounds. We investigated the changes of EMT markers and potential signaling pathways associated with the antitumor effect [13].

Three DNA methyltransferase inhibitors (DNMTi), 5-Azacytidine (AZA), Decitabine (DAC), and Guadecitabine/SGI-110 (SGI), and six histone deacetylase inhibitors (HDACi), JNJ-26481585 (JNJ), Vorinostat (SAHA), Entinostat (MS275), SB939, LBH589, and Tubastatin A HCL (Tub), were used in our experimental approach (Table 5.1).

DNMT inhibitors and HDAC inhibitors can reprogram highly aggressive TNBC cells that have undergone EMT to a less aggressive state, as evidenced by reduced

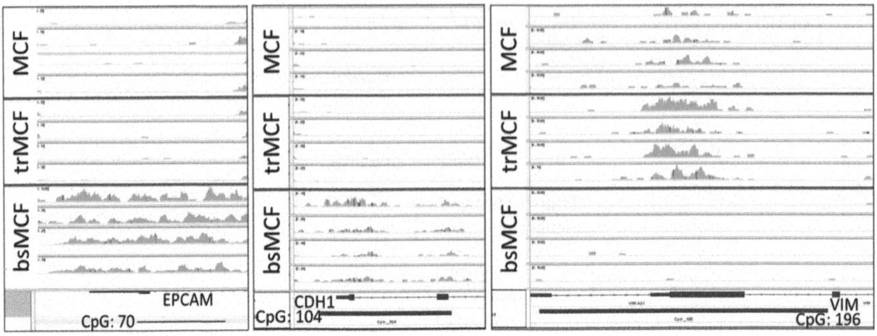

Fig. 5.5 Overall methylation on CpG islands of EMT hallmark genes. This hyper-methylation is affecting genes involved in biological functions crucial for the transformation process. Methylation profile was observed using Integrative Genomics Viewer, IGV (www. broadinstitute.org/igv/). Note that that EPCAM and CDH1 (E-cadherin), which are highly expressed in epithelial cells, are not methylated in MCF10F and trMCF; however both genes are methylated near the promoter in bsMCF cells. VIM (Vimentin), which is highly expressed in mesenchymal cells, presents methylation in MCF10F and trMCF, but not in bsMCF. (From Santucci-Pereira et al. [72])

Table 5.1 DNA methyltransferase and HDAC inhibitors

Drug name (abbreviation)	Provider (catalogue)	Drug category
5-Azacytidine (AZA)	Sigma (A2385)	DNMT inhibitor
5-Aza-2′-deoxycytindine (DAC)	Sigma (A3656)	DNMT inhibitor
SGI-110 sodium salt (SGI)	Astex (CK1463)	DNMT inhibitor
SGI-110 powder for injection	Astex	DNMT inhibitor
JNJ-26481585 (JNJ)	Selleckchem (S1096)	HDAC inhibitor
Vorinostat (SAHA)	Selleckchem (S1047)	HDAC inhibitor
Entinostat (MS275)	Selleckchem (S1053)	HDAC inhibitor
SB939	Selleckchem (S1515)	HDAC inhibitor
LBH589 (LBH)	Selleckchem (S1030)	HDAC inhibitor
Tubastatin A HCL (Tub)	Selleckchem (S2627)	HDAC inhibitor

cell proliferation, motility, and invasion, as well as colony formation (Figs. 5.6, 5.7, and 5.10).

The growth-inhibitory activity was assessed by MTT assay. All compounds induced a dose-dependent inhibition of proliferation, with the most sensitive to LBH and the least sensitive to Tub. The sensitivity of bsMCF cells to the three DNMTi was comparable. For comparison, both DAC and SGI were chosen for further studies. In contrast, the growth inhibitory effect in bsMCF cell line from six HDACi varied significantly. SGI but not DAC significantly inhibited migration of bsMCF cells (Wound closed: 66.9% ± 6.95% for control vs. 50.1% ± 4.2% for SGI over 16 hours). Four out of six HDACi significantly inhibited bsMCF cells migration. Among them, MS275 showed the greatest inhibition (Wound closed: 92.3% ± 7.6% for control vs. 55.2% ± 2.6% for MS275, 72.9% ± 4.5% for SB939, 77.9% ± 5.8% for LBH, 82.9% ± 3.7% for Tub over 18 hours) (Fig. 5.6).

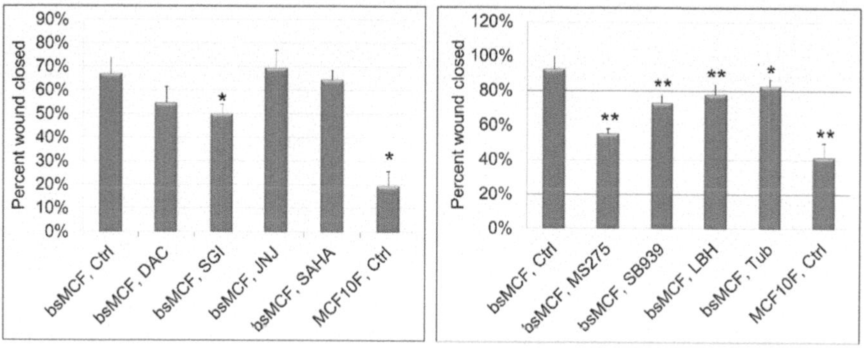

Fig. 5.6 Quantification of wound healing assay. *indicates $p < 0.05$; **indicates $p < 0.01$ compared to control. (Reprinted with permission of Su et al. [13])

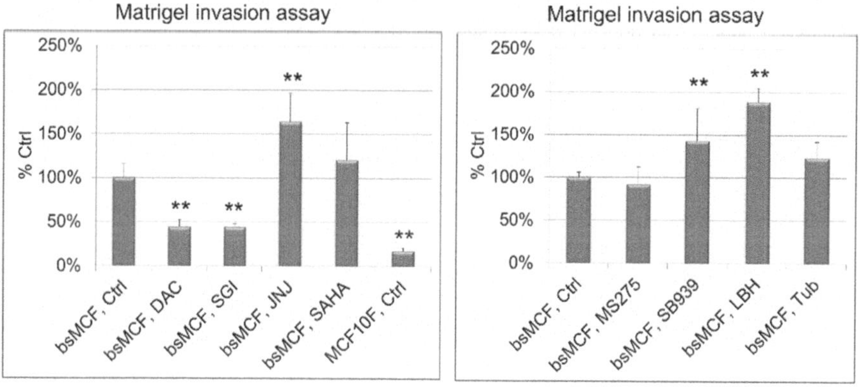

Fig. 5.7 Quantification of Matrigel invasion assay. *indicates $p < 0.05$; **indicates $p < 0.01$ compared to control. (Reprinted with permission of Su et al. [13])

Both DAC and SGI significantly suppressed bsMCF cell invasion by 56%. SAHA, MS275, and Tub did not have much effect. Interestingly, although the growth of bsMCF cells was significantly inhibited by JNJ, SB939, and LBH at very low doses treatment, the three drugs significantly increased cell invasion by 64% (JNJ), 43% (SB939), and 88% (LBH) (Fig. 5.7), indicating that the inhibition of cell growth should not be used as the only parameter for selecting the drug for cancer treatment. Drugs like JNJ, SB939, and LBH can promote cell invasion while inhibiting cell growth, thus may accelerate the progression of the disease.

The 3D cell culture is a technique that more closely resembles the cell growth behavior in vivo. When cultured in 3D collagen matrix, MCF10F cells form branching ductal structures. Tumorigenic bsMCF cell line usually does not form ductal structures; instead, it forms solid mass displaying protrusions and shows disseminated growth of individual cells [12]. The number of solid masses may indicate the potential of tumor cells to form a structure mimicking the primary tumor with

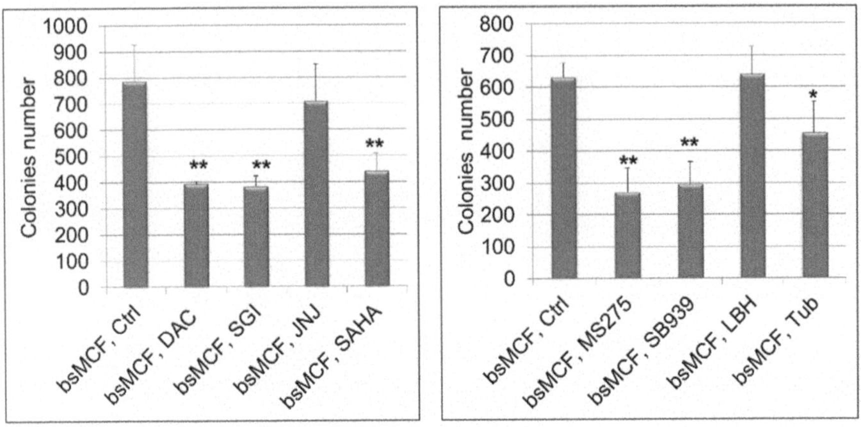

Fig. 5.10 Quantification of the number of the colonies formed in methylcellulose matrix. * indicates $p < 0.05$. ** indicates $p < 0.01$ compared to control. (Reprinted with permission of Su et al. [13])

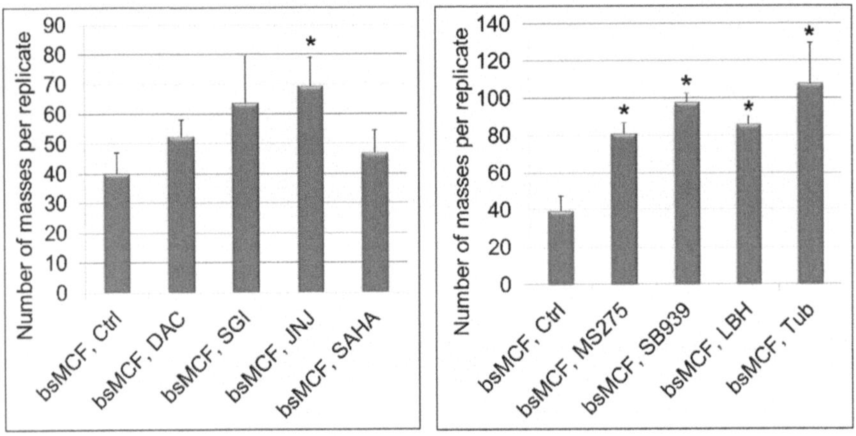

Fig. 5.8 DNMTi and HDACi reverse bsMCF cells malignant behavior in collagen. Bar graph shows quantification of the number of masses formed in collagen matrix. *indicates $p < 0.05$ compared to control. (Reprinted with permission of Su et al. [13])

cell-cell contact, whereas the protrusions resemble cell invasion, and the disseminated cell growth exhibits the mesenchymal feature of the tumor cells.

Treatment of JNJ, MS275, SB939, LBH, and Tub significantly increased the number of masses. DAC, SGI, and SAHA treatment also showed a trend of increasing the masses number (Fig. 5.8), suggesting the treatment increases cell-cell contact. The masses were then classified based upon the display of protrusions. DAC, SGI, and MS275 significantly increased type 1 masses, which showed no protrusions on the outer surface of the mass (Fig. 5.9). SGI is superior to DAC, and MS275 is superior to other HDACi with regard to the formation of type 1 masses, implying

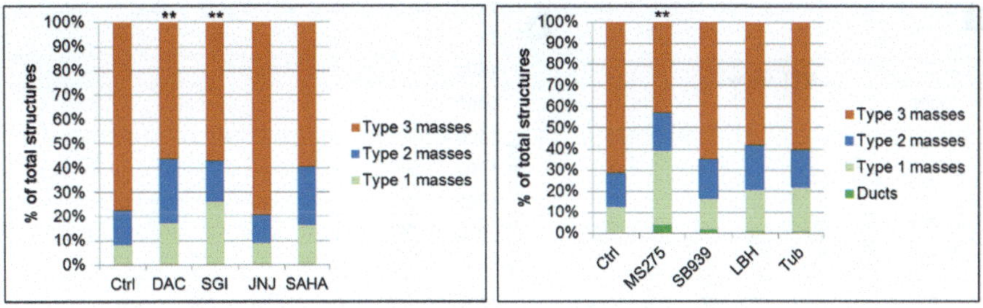

Fig. 5.9 Quantification of the types of the masses formed in collagen matrix. DAC, SGI, and MS275 treatment increase the number of masses with fewer protrusions. Type 1 masses show smooth outer surface and no protrusions present; type 2 masses show rough outer surface with 1–5 protrusions; type 3 masses show large amount of protrusions (>5), indicating more invasive phenotype. ** indicates $p < 0.01$ compared to control. (Reprinted with permission of Su et al. [13])

cells treated with SGI or MS275 are less invasive. It is noteworthy that DAC, SGI, and MS275 treatment reduced the disseminated growth of the individual cells, suggesting the reversal of mesenchymal phenotype.

Anchorage-independent growth of cells is usually correlated to the tumorigenicity of transformed cells. We assessed the effect of drug treatment on the colony formation ability of bsMCF cells in methylcellulose matrix. DAC, SGI, SAHA, MS275, SB939, and Tub significantly reduced the colonies number formed by bsMCF cells. DAC and SGI were comparable in the extent of inhibition. MS275 was the most potent one among six HDACi in decreasing the colonies number (Fig. 5.10).

The treatment of DNMTi in combination with HDACi in cancers has been shown more effective than the treatment of single agent in some studies [28, 29]. Therefore, we evaluated the effect of combined treatment using bsMCF cells. Treatment of SGI combined with SAHA (SGI + SAHA) or with MS275 (SGI + MS275) induced a greater inhibition in cell proliferation compared to single agent (Fig. 5.11a). In addition, colony formation assay also showed SGI + SAHA and SGI + MS275 formed fewer colonies compared to SAHA or MS275 treatment alone. SGI + MS275 is better than SGI + SAHA in inhibiting both cell proliferation and colony formation (Fig. 5.11b); thus SGI + MS275 was chosen for further study. The effect of the combined treatment on cell migration and invasion was evaluated by RTCA assay at two different doses. A significant decrease of cell migration was observed by SGI + MS275 but not by single agent at low-dose treatment. At high dose, the SGI + MS275 induced a significant decrease in both migration and invasion compared to control or single agent alone (Figs. 5.12 and 5.13).

When examined cell growth in collagen matrix, the treatment of SGI + MS275 formed more masses than SGI treatment alone, although there was no difference in the type of masses between these two groups. By contrast, there was no significant difference in the number of masses between the SGI + MS275 and MS275 treatment, whereas SGI + MS275 increased the proportion of masses with smooth outer

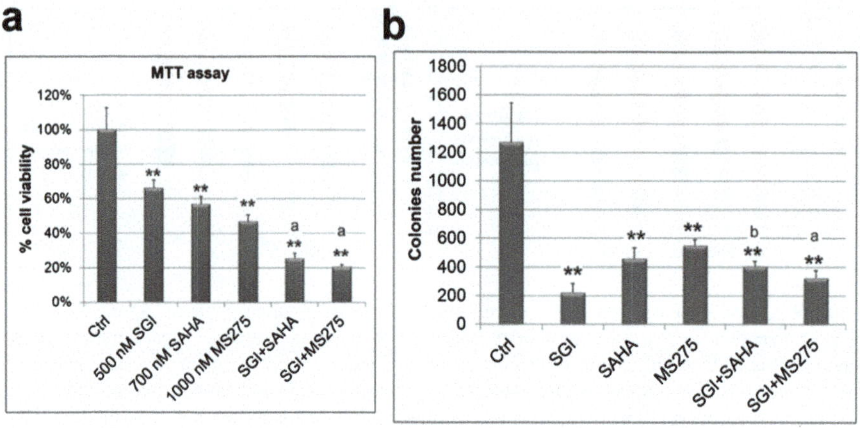

Fig. 5.11 Combined treatment of DNMTi with HDACi is more effective in bsMCF cells. (**a**) Combined treatment of SGI with SAHA or MS275 induces greater inhibition in cell proliferation. (**b**) Quantification of the number of colonies formed in agar methylcellulose matrix. The symbols used for indicating statistical results are as follows: *indicates $p < 0.05$ when compared to control. ** indicates $p < 0.01$ when compared to control. a indicates $p < 0.01$ when compared to the combined treatment with single agent. b indicates $p < 0.05$ when compared to the combined treatment with SGI. c indicates $p < 0.05$ when compared to the combined treatment with MS275. This rule applies to other figures for combined treatment too. (Reprinted with permission of: Su et al. [13])

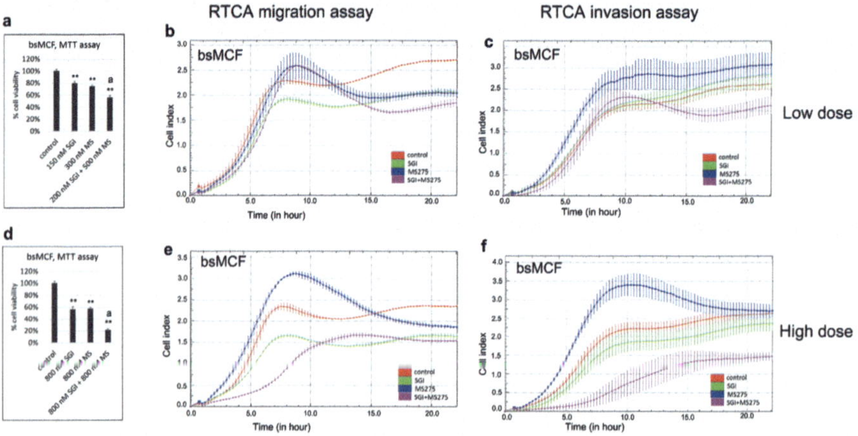

Fig. 5.12 Effects of SGI, MS275, and the combination on cell migration and invasion of bsMCF cells. (**a, d**) The doses used for migration and invasion assay were first evaluated by MTT assay to reveal the growth inhibition. (**b, e**) Cell index curve for RTCA migration with low- and high-dose treatment. (**c, f**) Cell index cure for RTCA invasion with low and high dose. Cells were treated with the dose shown in **a** & **d**; 4 days after the treatment, cells were trypsinized, viable cells were seeded in RTCA plate, and data were recorded every 15 minutes. (Reprinted with permission of Su et al. [13])

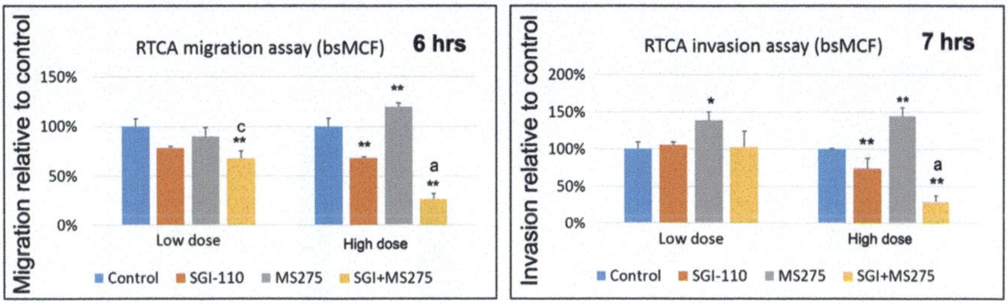

Fig. 5.13 Quantification graphs at the time point of 6 hours for RTCA cell migration, and 7 hours for RTCA invasion assay. Cells were first treated with the low or high dose indicated in Fig. 5.12, and then viable cells were used for both migration and invasion assay. (Reprinted with permission of Su et al. [13])

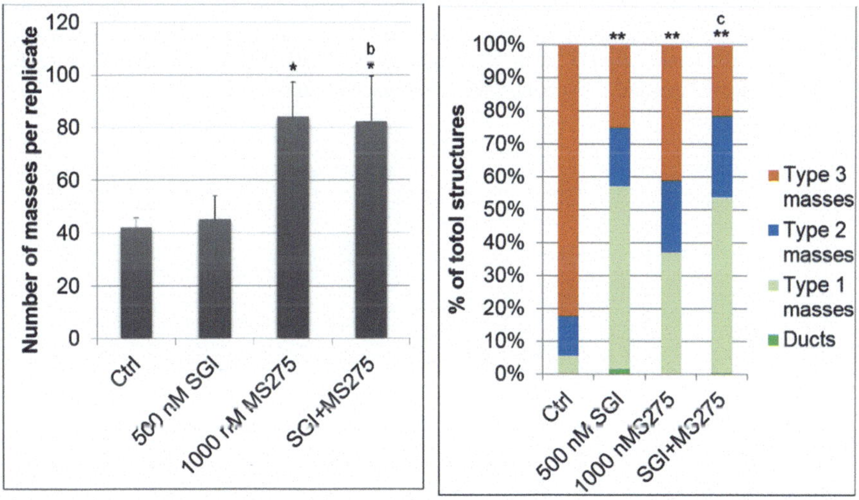

Fig. 5.14 Quantification of the number and types of masses formed by bsMCF cells in collagen matrix. (Reprinted with permission of Su et al. [13])

surface than MS275 treatment alone (Fig. 5.14), indicating the combined treatment is better than single agent in certain characteristics.

As floating cells were observed when treated with MS275 or SGI + MS275, the cell viability was assessed using trypan blue exclusion assay. MS275 induced 19.6% ± 6.0% and 19.5% ± 5.4% cell death in trMCF and bsMCF cells, respectively. The cell death was 23.8% ± 4.8% in trMCF and 23.3% ± 7.8% in bsMCF cells when treated with SGI + MS275. Although the proportion of cell death was similar between trMCF and bsMCF cells, the staining with cleaved caspase 3 showed the apoptosis was more pronounced in bsMCF cells than in trMCF cells. In addition, the combined treatment was more efficient in inducing apoptosis than single drug treatment (Fig. 5.15).

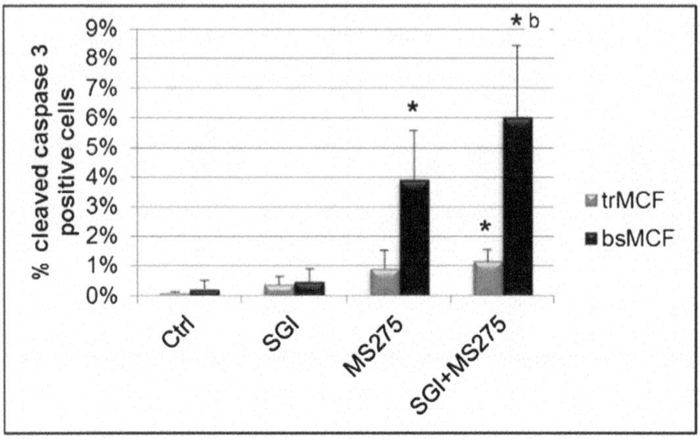

Fig. 5.15 MS275 alone and SGI+MS275 significantly induced apoptosis of trMCF and bsMCF cells evaluated by immunostaining with cleaved caspase 3. Graphs show the quantification of cells positive for cleaved caspase 3. (Reprinted with permission of Su et al. [13])

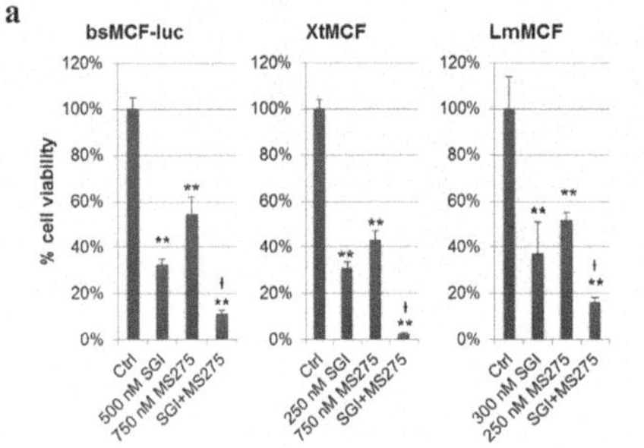

b

IC50 value of SGI and MS275 on the vability of cells

	bsMCF-luc	XtMCF	LmMCF
SGI (nM)	470±64	240±14	378±52
MS275 (nM)	1025±247	705±17	265±21

The concentrations used for *in vitro* studies

	bsMCF-luc	XtMCF	LmMCF
SGI (nM)	300	100	200
MS275 (nM)	750	500	250

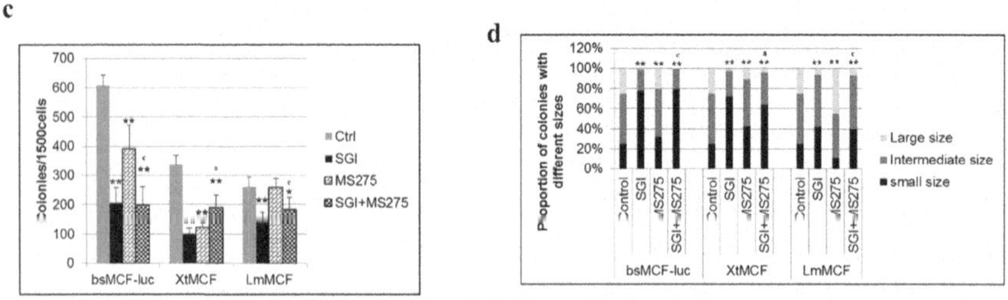

Fig. 5.16 Effect of SGI and MS275 in bsMCF-luc, XtMCF, and LmMCF cell lines. (**a**) Cells were plated in 96-well plate, treated on the following day, and then cultured for 96 hours; cell viability was measured by MTT assay. (b) Tables show IC50 of SGI and MS275 in the three cell lines, and the doses used for in vitro phenotypic study. (**c**, **d**) Quantification of the number and size of colonies formed in agar methylcellulose matrix. (Reprinted with permission of Su et al. [13])

The finding that SGI, MS275, and the combination exhibited in vitro antitumor activity against bsMCF cells prompted us to investigate whether these treatments have in vivo antitumor activity. For this purpose, we used two highly tumorigenic and metastatic TNBC cell lines XtMCF and LmMCF [12]. We first evaluated the in vitro antitumor activity. Compared to parental bsMCF-luc cells, the proliferation of XtMCF and LmMCF cells was more sensitive to SGI or MS275 (Fig. 5.16a). The combined treatment significantly reduced cell viability compared to single agent (Fig. 5.16a). Concentrations lower than IC50, as indicated in Fig. 5.16b, were then used for in vitro phenotypic studies. The colony formation assay (Fig. 5.16c) showed that SGI and SGI + MS275 induced a marked decrease of colony number in all three cell lines, whereas MS275 decreased colony number in bsMCF-luc and XtMCF

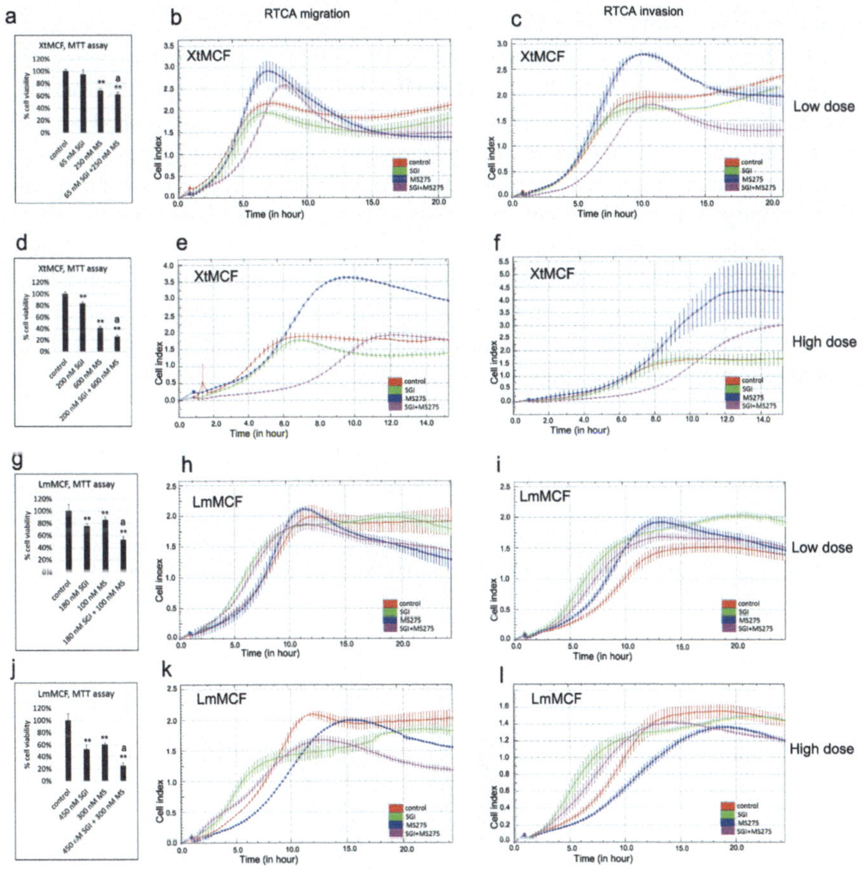

Fig. 5.17 Effects of SGI, MS275, and SGI+MS275 on cell migration and invasion of XtMCF and LmMCF cells. (**a**, **d**, **g**, **j**) The doses used for migration and invasion assay were first evaluated by MTT assay to reveal the growth inhibition. (**b**, **e**, **h**, **k**) Cell index curve for RTCA migration with low- and high-dose treatment. (**c**, **f**, **i**, **l**) Cell index curve for RTCA invasion with low- and high-dose treatment. (Reprinted with permission of Su et al. [13])

cells only. In addition, the size of the colonies was also reduced by SGI + MS275 in all three cell lines (Fig. 5.16d).

RTCA assay was used to evaluate the effect of the treatment on migration and invasion. The doses are indicated in Fig. 5.17a, d, g, j. For XtMCF cells, SGI treatment significantly inhibited cell migration (Fig. 5.17b, e) but not invasion (Fig. 5.17c, f). MS275 did not have much effect at any dose tested. Strikingly, SGI + MS275 significantly inhibited both migration and invasion at both low and high dose (Fig. 5.17b, c, e, f). LmMCF cells are very different from bsMCF and XtMCF cell; they are not invasive without adding additional growth factors in in vitro study. With basic FGF as chemoattractant, only high-dose MS275 showed inhibition in cell migration and invasion for LmMCF cell line, whereas SGI-110 or the SGI + MS275 promote both migration and invasion at low or high dose; one possible explanation is the change of growth factor receptor by the treatment may have influence on migration and invasion (Fig. 5.17h, i, k, l).

It has been reported the EMT generates cells with properties of stem cells [30]. We observed bsMCF-luc, XtMCF, and LmMCF cells have undergone EMT and are able to generate tumorspheres [12]. Tumorsphere formation assay showed SGI treated cells formed tumorspheres at the beginning in all three cell lines; however, the tumorspheres started to dissociate after 4 days culture. This phenomenon was more striking in bsMCF-luc cell line than in other two cell lines. Treatment of MS275 alone did not significantly affect the formation of tumorspheres. SGI + MS275 resulted in dissociation of almost all tumorspheres formed by bsMCF-luc cells, reduced mammospheres formed by XtMCF compared to MS275 alone, and significantly reduced the number of tumorspheres formed by LmMCF cells compared to single agent (Fig. 5.18).

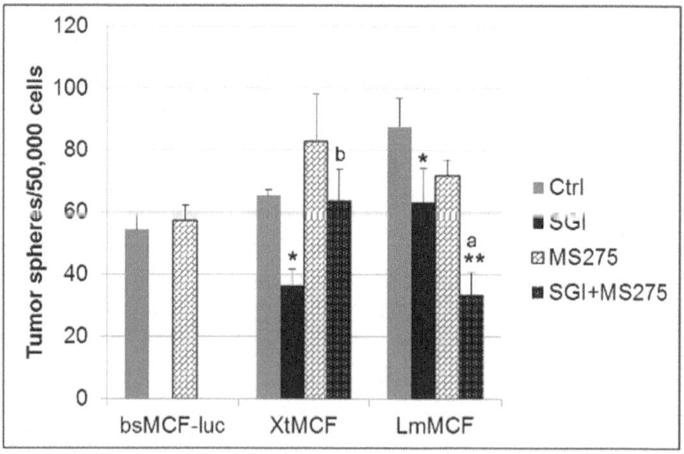

Fig. 5.18 Effects of SGI, MS275, or the combined treatment on tumorspheres formation by bsMCF-luc, XtMCF, and LmMCF cell lines. Graphs show the quantification of tumorsphere number. The tumorspheres formed by bsMCF-luc cells after the treatment with SGI or the combination were dissociated automatically and not counted. (Reprinted with permission of Su et al. [13])

The above results show the combined treatment is better than single agent in majority of the parameters we evaluated, and this effect is mainly dependent on the cell context, suggesting the importance of patient selection in clinics.

To evaluate the effect of the treatments on the in vivo growth of TNBC cells, 5×10^4 XtMCF cells were injected to CB17/SCID mice mammary fat pad. Three days after cells injection, mice were randomly divided into four groups and the treatment were initiated. All xenograft tumors continued growing during the course of therapy; however, the growth of the majority of tumors from treated animals was slower than tumors of control group in the week following the termination of therapy and thereafter. Mice were sacrificed 10 days after treatment stopped, there was a 54% decrease in the median tumor weight in MS275-treated group (83.5 mg in control vs. 38.5 mg in MS275), and a 52% decrease in SGI + MS275 group (83.5 mg

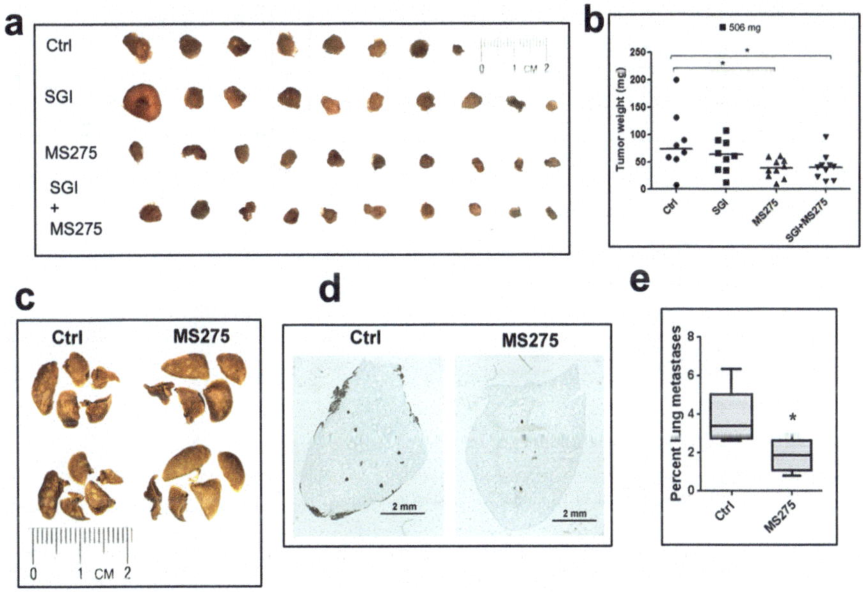

Fig. 5.19 MS275, or the combination of SGI with MS275, suppress TNBC growth or metastasis. (**a**) 5×10^4 XtMCF cells were injected to the mammary fat pad of 8–9-weeks-old female CB17/SCID mice. Three days after cells injection, mice were treated with indicted compounds for 3 weeks. The four treatment groups are (1) vehicle control; (2) SGI, 1.5 mg/kg body weight, Monday and Wednesday, s.c. injection; (3) MS275, 0.5 mg/kg body weight, Thursday, i.p. injection; and (4) SGI + MS275. Mice were sacrificed 31 days post cells injection. Images of XtMCF xenografts are shown. Magnification, 6.3×. (**b**) Dot plot shows the tumor weight at sacrifice; the line in each group indicates the median value. (**c**) 8×10^4 LmMCF cells were injected into the tail vein of 8–9-weeks-old female CB17/SCID mice. Six days after cells injection, mice were treated with vehicle control or 0.5 mg/kg body weight MS275 once a week for 3 weeks. Mice were sacrificed 25 days post cells injection. Representative images of lungs fixed with Bouin's solution show lung metastases on lung surface. Magnification, 8×. (**d**) Immunohistochemical staining to vimentin on lung sections show metastases of LmMCF cells in the lungs. Scale bar, 2 mm. (**e**) Quantification of the ratio of metastatic area to the whole lungs shows MS275 treatment inhibiting lung metastases of LmMCF cells. (Reprinted with permission of Su et al. [13])

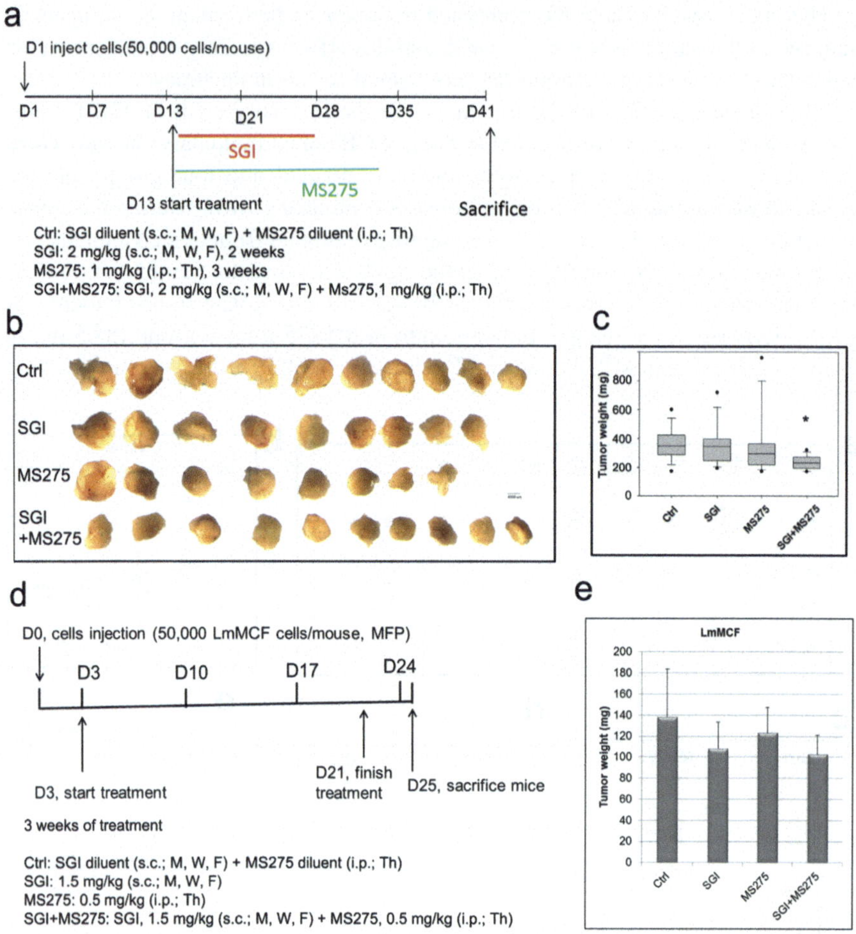

Fig. 5.20 Treatment of SGI, MS275, or the combination in xenograft model. (**a**) Treatment schedule for XtMCF xenograft in NOD/SCID mice. (**b**) Images of XtMCF xenografts are shown. Magnification, 6.3×. Scale bar: 4000 μm. (**c**) Box plots of the xenograft weight of XtMCF cells at sacrifice. Combined treatment significantly suppresses xenograft growth. (**d**) Treatment schedule for LmMCF xenograft in CB17/SCID mice. e Measurement of LmMCF xenograft weight. * indicates $p < 0.05$ when compared to control. (Reprinted with permission of Su et al. [13])

in control vs. 40 mg in SGI + MS275) (Fig. 5.19a, b). We also evaluated the effect of the treatments using NOD/SCID mice, with higher doses but a short treatment period. By this schedule, SGI+MS275 treatment reduced XtMCF tumor weight by 33% compared to control, although SGI or MS275 alone didn't show much effect on the tumor growth (Fig. 5.20a–c). Using LmMCF cells and CB17/SCID mouse model, all treated groups showed the trend of decrease in tumor weight but it didn't reach statistical difference (Fig. 5.20d, e). These results suggest that epigenetic drugs can exert a significant anti-tumor effect for TNBC in vivo when appropriate treatment schedule is used.

The lung metastatic study was conducted using LmMCF cell line; 80,000 cells were injected into the tail vein of CB17/SCID mice. Six days later, mice received MS275 by intra-peritoneal injection at 0.5 mg/kg once a week for 3 weeks and were sacrificed 25 days post cells injection. MS275 treatment reduced the tumor size on lungs surface compared to control shown by Bouin's solution fixation (Fig. 5.19c). The IHC staining to human vimentin on lung section showed the presence of tumor foci in the lungs (Fig. 5.19d); quantification of the lung areas affected by the metastases showed MS275 reduced lung metastases by twofold compared to control (1.9% ± 0.9% for MS275 vs. 3.8% ± 1.5% for control) (Fig. 5.19e). This result was consistent to the in vitro study that showed MS275 inhibits the migration and invasion of LmMCF cells (Fig. 5.17).

5.4 The Molecular Interpretation of How Epigenetic Reprograming May Affect the Reversion of the EMT Phenotype in the TNBC Cell Model

To investigate the mechanisms underlying the reversal of mesenchymal-like features induced by the treatment of DNMTi and HDACi, Western blot analyses were performed to assess the expression of EMT markers. All treatments induced upregulation of E-cadherin in bsMCF cells, without significant change of vimentin and SLUG expression (Fig. 5.21a).

The change of E-cadherin partially explained the reversal of bsMCF mesenchymal phenotype, but it could not address why the treatments of JNJ, SB939, and LBH promotes bsMCF cell invasion while upregulating E-cadherin. Our previous study suggested that the cleavage of EGF-like domain and the nuclear translocation of EpCAM are associated with EMT process [12]. Therefore, we examined the expression of EpCAM with antibodies targeting different epitopes (Fig. 5.22a, b). The staining using antibody EpCAM(VU1D9) targeting N-terminal domain showed there was no detectable EpCAM in bsMCF cells. However, using antibody EpCAM{Abbiotech} which recognizes thyroglobulin repeat-like domain and a part of cysteine poor region, we were able to observe the expression of EpCAM in these cells. The quantification of immunofluorescence showed that the nuclear EpCAM intensity was reduced in bsMCF cells treated with DAC, SGI, SAHA, MS275, and Tub; in contrast it was increased in bsMCF cells treated with JNJ and LBH (Fig. 5.21b, c), suggesting the elevated nuclear EpCAM level may be related to the increased cell invasion in JNJ- and LBH-treated bsMCF cells. EpCAM nuclear translocation is closely related to the WNT signaling; the intracellular domain of EpCAM can form a nuclear protein complex with beta-catenin, leading to gene transcription [31, 32]. Therefore, TCF4, the main effector of WNT signaling, was examined. All treatments reduced the expression of TCF4, and the change was more pronounced with the treatment of MS275 and SB939 (Fig. 5.21a). These results suggest both EpCAM and WNT signaling may contribute to the reversal of

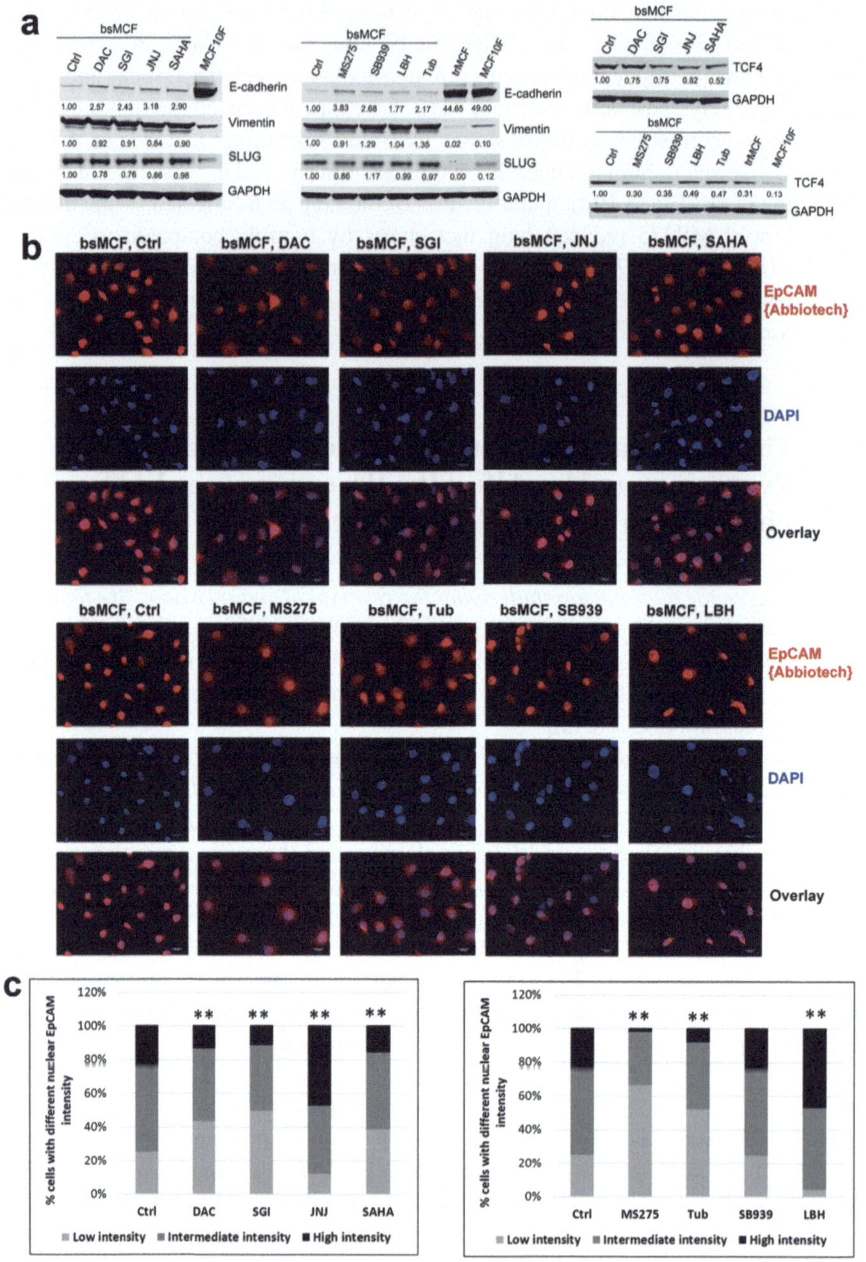

Fig. 5.21 Treatment of DNMTi and HDACi reprograms EMT partially mediated by EpCAM and WNT signaling. (**a**) Western blotting shows upregulation of E-cadherin and downregulation of TCF4 by the treatment of DNMTi and HDACi. The number below the band indicates the relative expression level to the control quantified by Ly-Cor Odyssey software. (**b**) Representative images of immunofluorescence staining of EpCAM. Magnification, 400×. Scale bar, 20 μm. (**c**) Quantification of nuclear EpCAM shows reduced nuclear EpCAM intensity in the cells treated with DAC, SGI, SAHA, MS275, and Tub, whereas an increased intensity was observed in the cells treated with JNJ and LBH. ** indicates $p < 0.01$ when compared to control. (Reprinted with permission of Su et al. [13])

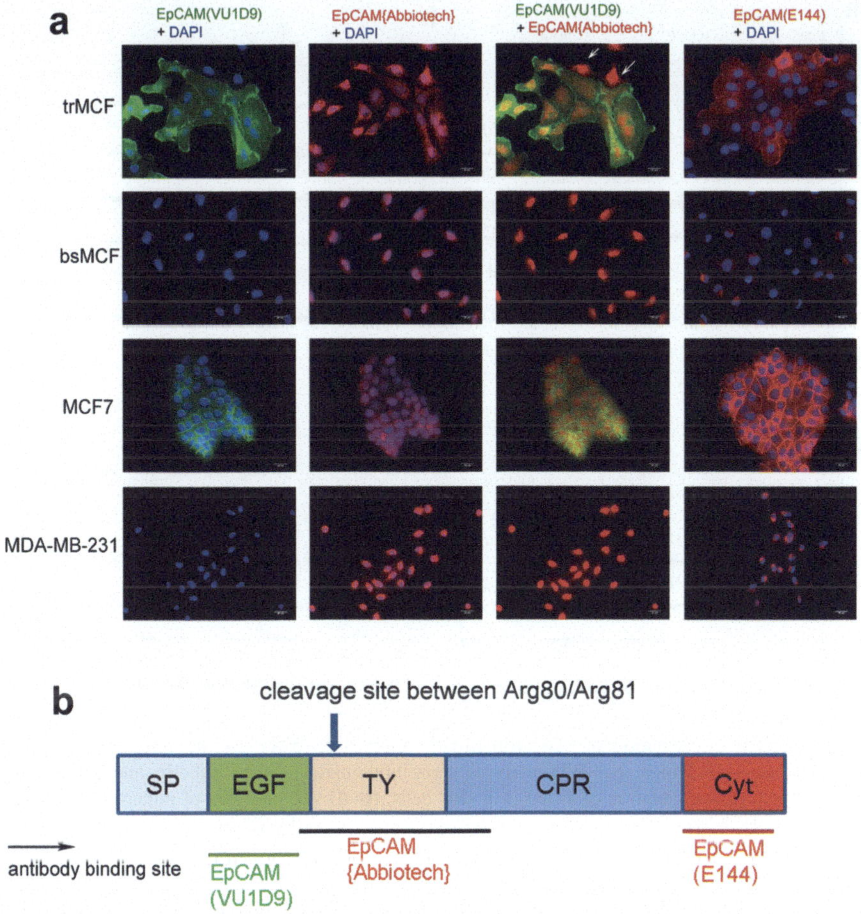

Fig. 5.22 N-terminal EGF-like domain of EpCAM is cleaved off after cells underwent EMT. (**a**) Immunofluorescence shows the staining of EpCAM in epithelial cell lines trMCF and MCF7, and mesenchymal-like cells bsMCF and MDA-MB-231 with antibodies to different epitopes. White arrows indicate trMCF cells that lost EGF-like domain showing a strong cytoplasmic and nuclear staining using antibody EpCAM{Abbiotech}. Scale bar, 20 μm (400×). (**b**) Schematic representation of EpCAM molecule and the epitopes of the antibodies used in this study. SP signal peptide, EGF EGF-like domain, TY thyroglobulin repeat-like domain, CPR cysteine poor region, Cyt cytoplasmic domain. (Reprinted with permission of Su et al. [13])

mesenchymal phenotype of bsMCF cells, but the crosstalk between these two pathways may be affected by the particular drug being used.

We next determined whether the combined treatment of SGI with MS275 would be superior to the single agents in regulating this signaling. Figure 5.23a shows E-cadherin was significantly upregulated in all three cell lines treated with the combination, whereas it was not detectable in control cells by Western blotting, and was just slightly observed in MS275-treated cells. Similarly, the full length EpCAM

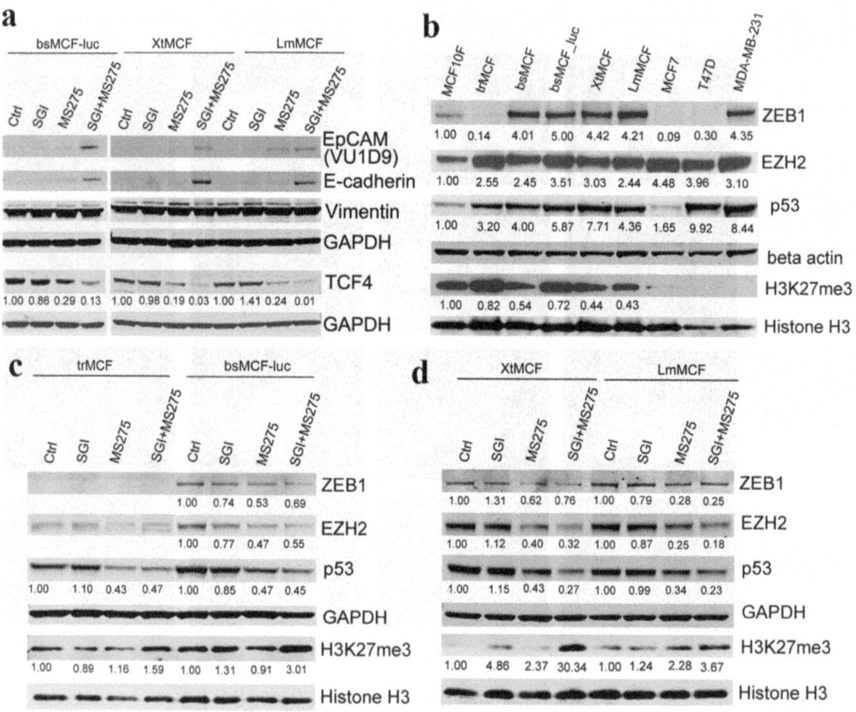

Fig. 5.23 The combination of SGI with MS275 suppresses WNT and mutant p53 pathways, upregulating H3K27me3. (**a**) WB analyses show upregulation of E-cadherin and full length EpCAM and downregulation of TCF4 in TNBC cells. (**b–d**) WB analyses of mutant p53, ZEB1, EZH2, and H3K27me3. Beta actin and GAPDH were used as loading control. (Reprinted with permission of Su et al. [13])

shown by the reactivity to the antibody EpCAM(VU1D9) was significantly increased in cells treated with the combination (Fig. 5.23a), and the fluorescence staining further proved the presence of full length EpCAM in these cells (Fig. 5.24), indicating the combined treatment protects EpCAM from being cleaved. Consistently, TCF4 was significantly reduced in cells treated with the combination when compared with the single agent (Fig. 5.23a). These findings suggest that SGI and MS275 have a synergistic effect on upregulating E-cadherin, and this effect may be partially mediated by inhibiting EpCAM cleavage and WNT signaling.

It has been reported that SAHA and MS275 suppress mutant p53 expression in TNBC and pancreatic cancer [33, 34]. *TP53* mutations are the most frequent genetic alterations in breast cancers, especially in basal-like cancers [35]. The mutant p53 can induce ZEB1 and EZH2 expression, thus promoting EMT [36, 37]. We performed cDNA sequencing of exon 5–8 (central DNA binding domain) of *TP53* gene following the literature [38] and found the tumorigenic cell lines bsMCF, bsMCF-luc, XtMCF, and LmMCF all harbor one copy of mutation in codon 241 (TCC in wild type vs. CCC in mutant); in contrast, the parental cell line MCF10F and trMCF

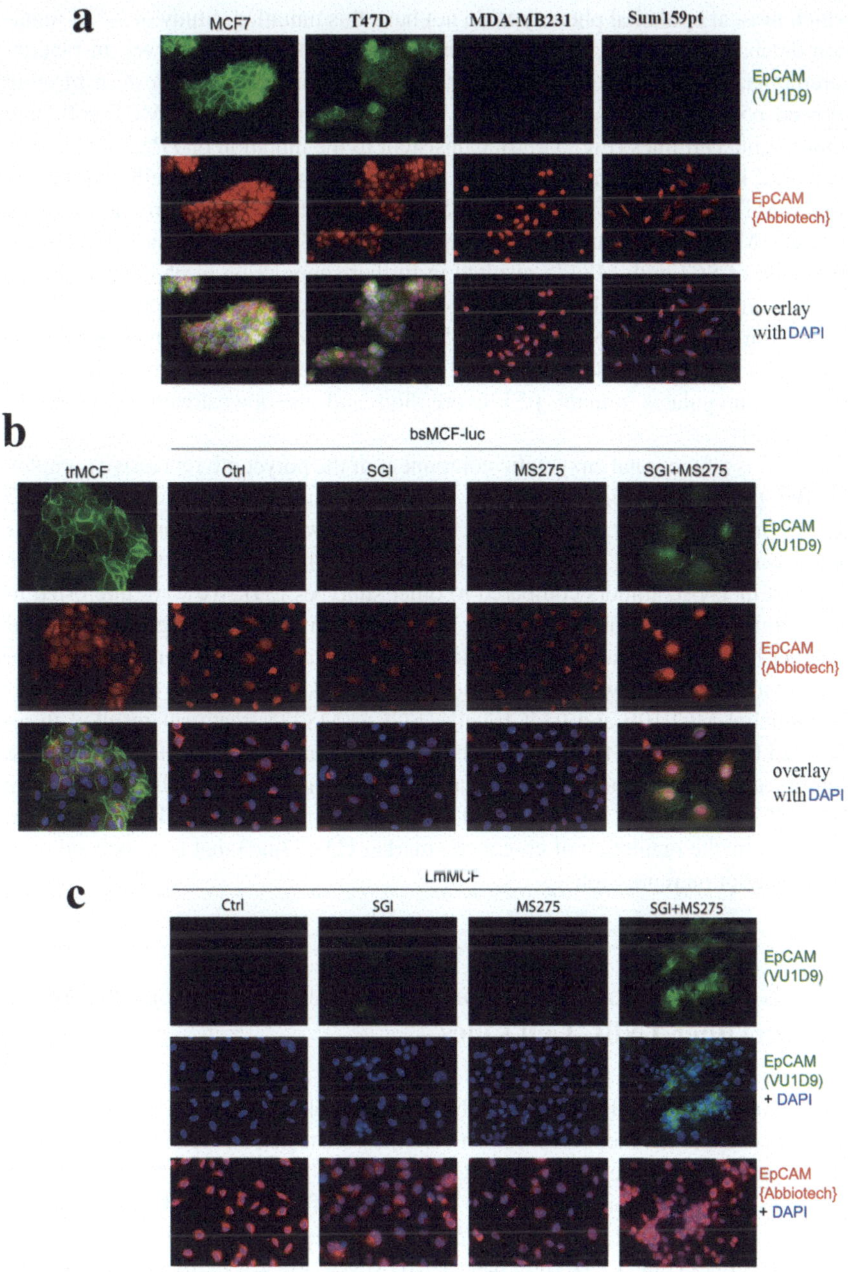

Fig. 5.24 Immunofluorescence staining of cells treated with single or combined agent. (**a**) Staining of EpCAM in luminal cell lines MCF7 and T47D, and mesenchymal-like basal cell lines MDA-MB231 and Sum159pt. (**b**, **c**) The combined treatment of SGI with MS275 upregulates full length EpCAM expression in bsMCF-luc and LmMCF cells. Cells were plated onto chamber slides and treated with indicated compounds on the following day, then cultured for 4 days and followed with the immunofluorescence staining. Scale bar, 20 μm (400×). (Reprinted with permission of Su et al. [13])

which present epithelial phenotype do not have this mutation. Study of *TP53* mutation database showed that the mutation in codon 241 is also observed in bladder carcinoma, brain tumor, breast cancer, and other cancer types. Western blotting showed p53 protein was significantly increased in transformed trMCF cells and tumorigenic cell lines (Fig. 5.23b). Consistent to the mutation of *TP53*, EZH2 was increased in all transformed cells, and ZEB1 was upregulated in cells that underwent EMT (Fig. 5.23b). Treatment of SGI did not change the expression of p53, whereas MS275 treatment significantly reduced p53 in all cell lines. Combined treatment of SGI with MS275 resulted in further suppression in p53 expression in XtMCF and LmMCF cells compared to single agent. Similarly, EZH2 and ZEB1 expression were inhibited by MS275 alone or MS275 combined with SGI (Fig. 5.23c, d). Our results revealed that MS275, or the combination of MS275 with SGI, downregulates mutant p53 expression and its downstream signaling in TNBC cells.

EZH2 is a functional enzymatic component of the polycomb repressive complex 2. The primary function of EZH2 is to methylate lysine-27 on histone H3 (H3K27me3). However, EZH2 and H3K27me3 show an inverse correlation across breast cancer subtypes; the EZH2 expression is the highest and the H3K27me3 is the least in TNBC tumors compared to other subtypes [39]. We were interested in determining whether the H3K27me3 level was also altered during the EMT and whether it can be regulated by SGI and MS275 treatment. As shown in Fig. 5.23b, the global H3K27me3 was lower in mesenchymal-like tumorigenic cells compared to epithelial MCF10F and trMCF cells. SGI or MS275 treatment resulted in an elevated H3K27me3 in two highly tumorigenic cell lines XtMCF and LmMCF; the combined treatment synergistically increased H3K27me3 in all cell lines, especially in XtMCF (Fig. 5.23c, d). This result indicates the reversal of EMT may also be mediated by the regulation of epigenetic marker H3K27me3 that is associated with the transcription repression.

5.5 Generalization of the Concept Epigenetic Reprograming in Other TNBC Cell Lines

To determine whether the growth inhibitory effects of SGI, MS275, and the combination also apply to other breast cancer cell lines, we tested the treatments in a panel of commercial available luminal, HER2 type, and TNBC cell lines. The MCF series cell lines were used for comparison. The growth of luminal and HER2 type cell lines was not sensitive to SGI (Fig. 5.25a, middle panel). Among the six tested TNBC cell lines (Fig. 5.25a, right panel), MDA-MB-468 was extremely sensitive to SGI (IC50 of 20 nM). All tested cell lines showed a dose-dependent response upon MS275 treatment. The dose around IC50 was then used to do combined treatment. For the cell lines with the IC50 above 1000 nM, cells were treated with the agent at 1000 nM. Although luminal and HER2 type cell lines did not respond very well to

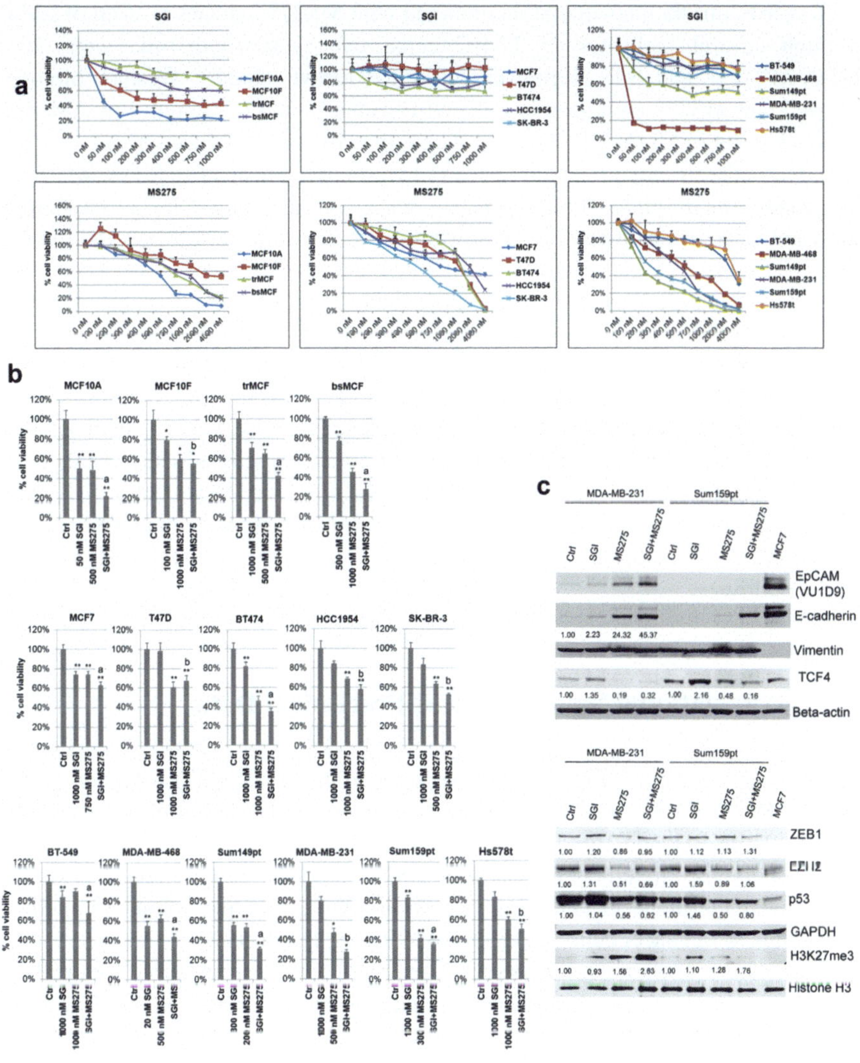

Fig. 5.25 TNBC cell lines are sensitive to the combined treatment of SGI with MS275. (**a**) Dose response curve of breast epithelial/cancer cell lines to the treatment of SGI and MS275 by MTT assay. (**b**) The effect of SGI, MS275, or the combination treatment on the growth of breast epithelial/cancer cell lines by MTT assay. (**c**) Western blotting shows inhibition of WNT and mutant p53 signaling pathways and upregulation of H3K27me3. (Reprinted with permission of Su et al. [13])

SGI alone at concentrations below 1000 nM, combined treatment showed greater inhibition in cell growth in MCF7, BT-474, HCC1954, and SK-BR-3 cell lines. The MCF series and all other TNBC cell lines showed significantly inhibitory effect upon the combined treatment, with the most sensitive in MDA-MB-468 and Sum149pt cell lines (Fig. 5.25b).

To verify our findings that SGI combined with MS275 induced reversal of EMT through several pathways in TNBC cells, two mesenchymal-like cell lines MDA-MB-231 and Sum159pt were used for further studies; Western blotting showed that MS275 significantly induced E-cadherin and full length EpCAM and reduced TCF4 in MDA-MB-231 cells. Combined treatment induced higher expression of E-cadherin and full length EpCAM than single agent alone in MDA-MB-231. The restoration of full length EpCAM was not observed in Sum159pt cells although the combined treatment showed significant upregulation of E-cadherin and inhibition of TCF4 (Fig. 5.25c). Both MDA-MB-231 and Sum159pt cells have *TP53* missense mutation [40]. Western blotting showed there was a notable decrease of p53 in both cell lines after MS275 and the combined treatment. EZH2 was also reduced by MS275 and the combined treatment in MDA-MB-231, but not in Sum159pt cells (Fig. 5.25c). There was no significant change in ZEB1 expression in both cell lines with any of the treatments. H3K27me3 was increased by MS275, with greater increase by the combined treatment in both cell lines. These results suggest that the growth inhibition and the reversal of EMT by the combined treatment of SGI with MS275 are not limited to our MCF series model, but also apply to other TNBC cells presenting EMT features; nevertheless the involved pathways might be dependent on the cell lines.

5.6 Translational Significance of Preclinical Data

The second generation DNMTi SGI is superior to the two DNMTi AZA and DAC, and the HDACi MS275 is more potent than other five HDAC inhibitors (SAHA, JNJ, SB939, LBH, and Tub) tested in this study. We demonstrated for the first time that the combined treatment of SGI with MS275 exerts a greater effect than single agent treatment alone in inhibiting cell proliferation, motility, colony formation, and stemness of cancer cells, as well as in inducing apoptosis.

The DNA methylation inhibitors AZA and DAC have been approved for medical use in the treatment of leukemia and myelodysplastic syndrome (MDS); however, the use in breast cancer is still under research [41, 42]. Clinical trials of using DAC before surgery in treating patient with locally advanced HER2-negative breast cancer and metastatic TNBC are ongoing. SGI, also known as guadecitabine, is a dinucleotide of a decitabine (DAC) linked via phosphodiester bond to a guanosine; this configuration provides protection from drug clearance by deamination, leading to prolonged exposure to decitabine. SGI exerts immunomodulatory action in leukemia, ovarian cancer, and breast cancer [43–45], sensitizing cancer cells to other agents [29, 46, 47]. SGI is currently used in clinical trials for some cancers including non-small lung cancer, recurrent ovarian cancer, metastatic colorectal cancer, hepatocellular carcinoma, MDS, and leukemia. The investigation of SGI in breast cancer and clinical trial has not been reported yet.

Use of HDAC inhibitors in breast cancer treatment is being extensively studied in recent years. Pan-HDAC inhibitor SAHA was the first HDAC inhibitor approved by the US Food and Drug Administration for the clinical use in 2006. SAHA is currently in phase II clinical trial for breast cancer patients. Three other pan-HDAC inhibitors, JNJ, LBH, and SB939, are studied for the treatment of different cancers including breast cancer. Tub is a selective HDAC6 inhibitor; it shows a dose-dependent effect on cell viability in MDA-MB-231 and induces apoptosis [48]. MS275 is a selective HDAC1 and HDAC3 inhibitor. MS275 induces apoptosis of erbB2-overexpressing cells [49], sensitizes TRAIL-resistant breast cancer cells, inhibits angiogenesis and metastasis, as well as reverses EMT [50, 51]. MS275 also has an effect in inhibiting tumor-initiating cells [52]. MS275 is currently used in multiple phase III clinical trials in treating breast cancer.

In this study we investigated the mechanism of action of DNMTi and HDACi in TNBC cells and showed that E-cadherin was significantly induced in mesenchymal-like cancer cells by all treatments, suggesting the reversal of EMT by epigenetic remodeling. E-cadherin is one of the most important molecules in epithelial cells; it interacts with a bundle of proteins and regulates basic cellular process. One protein that E-cadherin interacts with is beta-catenin. Research shows E-cadherin inhibits tumor cell growth by antagonizing beta-catenin signaling [53]. The upregulation of E-cadherin might contribute to the inhibition of cell growth by DNMTi and HDACi. Of course, both DNMTi and HDACi can also inhibit cell proliferation through other anti-tumor effects such as inducing cell cycle arrest [54, 55].

We also found that the expression pattern of EpCAM is closely related to the status of EMT and the invasion property of the cancer cells. The combined treatment of SGI with MS275 significantly increased the full length EpCAM protein, in a pattern that was parallel to the upregulation of E-cadherin, implying these cell membrane proteins are important to help cells stick to one another (cell adhesion). EpCAM has been considered as an oncogenic signal transducer; the cleaved EpCAM helps relay signals from outside of cells to nucleus, resulting to cross-talk to other signaling including WNT pathway [56]. In our study, we observed a decrease of TCF4 by DNMTi and HDACi in mesenchymal-like cell lines. It is of importance that the combined treatment of SGI with MS275 synergistically increases full-length EpCAM, inhibits TCF4, and upregulates E-cadherin. Based on these findings, we postulate that the cleavage of N-terminal EGF-like domain of EpCAM may cause the internalization of the cleaved EpCAM, which cross-talks with WNT signaling and other pathways, contributing to EMT process. The combined treatment of SGI with MS275 synergistically inhibits the cleavage of EpCAM, thus suppressing WNT signaling and reversing EMT. Given the fact that WNT/beta-catenin signaling is related to TNBC tumorigenesis and metastasis [57, 58], and there are no drugs targeting WNT pathways currently approved for the treatment of TNBC, our results strongly suggest combined treatment of SGI with MS275 could be a strategy targeting EMT through WNT pathway in TNBC.

In addition, we observed that MS275 alone or combined with SGI suppresses the expression of mutant p53 and reduces ZEB1 and EZH2 protein in mesenchymal-like TNBC cells. Loss of p53 function is associated with the induction of EMT and acquisition of stemness in different cancer cell lines [59, 60]. ZEB1 is one of the most important E-cadherin repressors in EMT process. ZEB1 can be induced by mutant p53 [36] or loss of wild-type p53 [60] through modulating miRNAs. EZH2 is a marker of aggressive breast cancer; overexpression of EZH2 promotes neoplastic transformation of breast epithelial cells [61]. Breast cancers with *TP53* mutation correlate with high EZH2 expression [62]. Jiang et al. demonstrated that mutant p53 induces EZH2 expression and promotes EMT via miR-26a in endometrial carcinoma cells [37]. Thus, the reversal of EMT by MS275 and the combination could be partially through the suppression of mutant p53. Furthermore, both mutant p53 and EZH2 have multiple functions in tumorigenesis other than inducing EMT [63, 64]. SGI combined with MS275 could also be a strategy targeting mutant p53 and EZH2 in TNBC treatment.

One interesting finding in this study is the observation of the significant increase of H3K27me3 level in treated cells. Breast cancer patients with low H3K27me3 have a shorter survival than those with intermediate and high levels [39]. H3K27me3 level is usually maintained by polycomb repressive complex and histone demethylases. KDM6B is a demethylase that specifically demethylates H3K27me3; it is required by TGF beta-induced EMT in mammary cells and is highly expressed in invasive breast cancer [65]. Wang et al. reported that miR-138 suppresses EMT through degradation of KDM6B in breast cancer carcinoma [66]. A more recent study showed that treatment of MCF7 and MDA-MB-231 cells with GSKJ4 inhibits self-renewal of cancer stem cells by inhibiting KDM6B and UTX (another histone demethylase) and elevating the global H3K27me3 level [67]. Based on these studies, we postulate that one of the reasons that causes low H3K27me3 in TNBC could be the overexpression of KDM6B, and the combined treatment of SGI with MS275 might probably regulate KDM6B or other histone demethylase through miRNA or other pathways, which result in the increase of H3K27me3 and chromatin remodeling, hence contributing to reversal of EMT.

In this work, although SGI or MS275 treatment alone showed almost similar effect in reversing EMT phenotype in vitro, the in vivo study using CB17/SCID mice showed that MS275 was more potent in suppressing xenograft growth. This could be due to the treatment schedule used for SGI being not optimal. We found that SGI was actually more effective than MS275 in inhibiting colony formation and stemness of TNBC cell lines. Pathania et al. reported that DNMT1 is essential for mammary and cancer stem cell maintenance and tumorigenesis [68]. Assessment of DNMT1 in breast cancer tissues showed that DNMT1 is highly expressed in TNBC compared with other subtypes [69]. DNMT1 is a direct target of SGI; treatment of SGI depletes DNMT1 in hepatocellular carcinoma cell line [47]. Moreover, DNMT1 is associated with TGF beta-induced EMT in ovarian cancer cells, and SGI treatment prevents this EMT process [70]. These findings suggest that SGI has the potential to reduce the risk of TNBC metastasis by inhibiting EMT and stemness of cancer cells.

Similar to the other chemotherapies, the use of DNMTi and HDACi in treating solid tumors was not very successful in clinics. Many reasons such as toxicity, poor penetration, drug resistance, and poor patient section could attribute to this failure. Compared to other DNMTi, SGI-110 has a prolonged half-life that allows to use it at a lower dose and longer interval, thus reducing the toxicity. MS275 is a selective

HDAC1 and HDAC3 inhibitor; it has multiple anti-tumor effects. In the present study, we focus on EMT only and observed that SGI combined with MS275 inhibits EMT in mesenchymal-like TNBC cells. We showed that TNBC cell lines are more sensitive to the treatment of SGI combined with MS275 compared with other subtypes, implying the clinical significance in patient selection. Taken together, our results clearly suggest that the combination of SGI with MS275 may have clinical advantage over other DNMTi and HDACi in treating TNBC.

5.7 Conclusions

Our studies showed that DNMTi and HDACi exert antitumor activity in TNBC cells by epigenetically reprograming EMT. We demonstrated that the DNMTi SGI and HDACi MS275 have synergistic effect on reversing EMT by inhibiting EpCAM cleavage and WNT signaling, suppressing mutant p53, ZEB1, and EZH2 expression, and inducing E-cadherin, apoptosis, as well as histone H3 lysine 27 tri-methylation. Our findings strongly suggest that TNBC is sensitive to epigenetic-based therapies. We propose a new strategy to treat TNBC by using the combination of SGI-110 with MS275, which exerts superior antitumor effects by simultaneously targeting multiple pathways that reverse EMT.

References

1. Masuda, H., Baggerly, K. A., Wang, Y., Zhang, Y., Gonzalez-Angulo, A. M., Meric-Bernstam, F., et al. (2013). Differential response to neoadjuvant chemotherapy among 7 triple-negative breast cancer molecular subtypes. *Clinical Cancer Research: An Official Journal of the American Association for Cancer Research, 19*(19), 5533–5540.
2. Andre, F., & Zielinski, C. C. (2012). Optimal strategies for the treatment of metastatic triple-negative breast cancer with currently approved agents. *Annals of Oncology: Official Journal of the European Society for Medical Oncology/ESMO, 23*(Suppl 6), vi46–vi51.
3. Crown, J., O'Shaughnessy, J., & Gullo, G. (2012). Emerging targeted therapies in triple negative breast cancer. *Annals of Oncology: Official Journal of the European Society for Medical Oncology/ESMO, 23*(Suppl 6), vi56–vi65.
4. Bayraktar, S., & Gluck, S. (2013). Molecularly targeted therapies for metastatic triple-negative breast cancer. *Breast Cancer Research and Treatment, 138*(1), 21–35.
5. Chaffer, C. L., & Weinberg, R. A. (2011). A perspective on cancer cell metastasis. *Science, 331*(6024), 1559–1564.
6. Valastyan, S., & Weinberg, R. A. (2011). Tumor metastasis: Molecular insights and evolving paradigms. *Cell, 147*(2), 275–292.
7. Liang, Q., Li, L., Zhang, J., Lei, Y., Wang, L., Liu, D. X., et al. (2013). CDK5 is essential for TGF-beta1-induced epithelial-mesenchymal transition and breast cancer progression. *Scientific Reports, 3*, 2932.
8. Pang, M., Georgoudaki, A., Lambut, L., Johansson, J., Tabor, V., Hagikura, K., et al. (2016). TGF-β1-induced EMT promotes targeted migration of breast cancer cells through the lymphatic system by the activation of CCR7/CCL21-mediated chemotaxis. *Oncogene, 35*(6), 748–760.

9. Yook, J. I., Li, X. Y., Ota, I., Hu, C., Kim, H. S., Kim, N. H., et al. (2006). A Wnt-Axin2-GSK3beta cascade regulates Snail1 activity in breast cancer cells. *Nature Cell Biology, 8*(12), 1398–1406.
10. May, C. D., Sphyris, N., Evans, K. W., Werden, S. J., Guo, W., & Mani, S. A. (2011). Epithelial-mesenchymal transition and cancer stem cells: A dangerously dynamic duo in breast cancer progression. *Breast Cancer Research, 13*(1), 202.
11. Morel, A.-P., Lièvre, M., Thomas, C., Hinkal, G., Ansieau, S., & Puisieux, A. (2008). Generation of breast cancer stem cells through epithelial-mesenchymal transition. *PLoS One, 3*(8), e2888.
12. Su, Y., Pogash, T. J., Nguyen, T. D., & Russo, J. (2016). Development and characterization of two human triple-negative breast cancer cell lines with highly tumorigenic and metastatic capabilities. *Cancer Medicine, 5*(3), 558–573. https://doi.org/10.1002/cam4.616. PMID: 26775583.
13. Su, Y., Hopfinger, N. R., Nguyen, T. D., Pogash, T. J., Santucci-Pereira, J., & Russo, J. (2018). Epigenetic reprogramming of epithelial mesenchymal transition in triple negative breast cancer cells with DNA methyltransferase and histone deacetylase inhibitors. *Journal of Experimental & Clinical Cancer Research: CR, 37*(1), 314. https://doi.org/10.1186/s13046-018-0988-8. PMID:30547810
14. Idowu, M. O., Kmieciak, M., Dumur, C., Burton, R. S., Grimes, M. M., Powers, C. N., et al. (2012). CD44(+)/CD24(−/low) cancer stem/progenitor cells are more abundant in triple-negative invasive breast carcinoma phenotype and are associated with poor outcome. *Human Pathology, 43*(3), 364–373.
15. Jang, M. H., Kim, H. J., Kim, E. J., Chung, Y. R., & SYJHP, P. (2015). Expression of epithelial-mesenchymal transition–related markers in triple-negative breast cancer: ZEB1 as a potential biomarker for poor clinical outcome. *Human Pathology, 46*(9), 1267–1274.
16. Peinado, H., Ballestar, E., Esteller, M., & Cano, A. (2004). Snail mediates E-cadherin repression by the recruitment of the Sin3A/histone deacetylase 1 (HDAC1)/HDAC2 complex. *Molecular and Cellular Biology, 24*(1), 306–319.
17. Hajra, K. M., Chen, D. Y., & Fearon, E. R. (2002). The SLUG zinc-finger protein represses E-cadherin in breast cancer. *Cancer Research, 62*(6), 1613–1618.
18. Park, S. M., Gaur, A. B., Lengyel, E., & Peter, M. E. (2008). The miR-200 family determines the epithelial phenotype of cancer cells by targeting the E-cadherin repressors ZEB1 and ZEB2. *Genes & Development, 22*(7), 894–907.
19. Vesuna, F., van Diest, P., Chen, J. H., & Raman, V. (2008). Twist is a transcriptional repressor of E-cadherin gene expression in breast cancer. *Biochemical and Biophysical Research Communications, 367*(2), 235–241.
20. Huang, Y., Fernandez, S. V., Goodwin, S., Russo, P. A., Russo, I. H., Sutter, T. R., et al. (2007). Epithelial to mesenchymal transition in human breast epithelial cells transformed by 17beta-estradiol. *Cancer Research, 67*(23), 11147–11157.
21. Lombaerts, M., Van Wezel, T., Philippo, K., Dierssen, J., Zimmerman, R., Oosting, J., et al. (2006). E-cadherin transcriptional downregulation by promoter methylation but not mutation is related to epithelial-to-mesenchymal transition in breast cancer cell lines. *British Journal of Cancer, 94*(5), 661–671.
22. Graff, J. R., Herman, J. G., Lapidus, R. G., Chopra, H., Xu, R., Jarrard, D. F., et al. (1995). E-cadherin expression is silenced by DNA hypermethylation in human breast and prostate carcinomas. *Cancer Research, 55*(22), 5195–5199.
23. Hemavathy, K., Guru, S. C., Harris, J., Chen, J. D., & Ip, Y. T. (2000). Human Slug is a repressor that localizes to sites of active transcription. *Molecular and Cellular Biology, 20*(14), 5087–5095.
24. Shi, Y., Sawada, J., Sui, G., Affar el, B., Whetstine, J. R., Lan, F., et al. (2003). Coordinated histone modifications mediated by a CtBP co-repressor complex. *Nature, 422*(6933), 735–738.

25. Aghdassi, A., Sendler, M., Guenther, A., Mayerle, J., Behn, C.-O., Heidecke, C.-D., et al. (2012). Recruitment of histone deacetylases HDAC1 and HDAC2 by the transcriptional repressor ZEB1 downregulates E-cadherin expression in pancreatic cancer. *Gut, 61*(3), 439–448.
26. Fu, J., Qin, L., He, T., Qin, J., Hong, J., Wong, J., et al. (2011). The TWIST/Mi2/NuRD protein complex and its essential role in cancer metastasis. *Cell Research, 21*(2), 275–289.
27. Soule, H. D., Maloney, T. M., Wolman, S. R., Peterson, W. D., Brenz, R., McGrath, C. M., et al. (1990). Isolation and characterization of a spontaneously immortalized human breast epithelial cell line, MCF-10. *Cancer Research, 50*(18), 6075–6086.
28. Fan, J., Yin, W.-J., Lu, J.-S., Wang, L., Wu, J., Wu, F.-Y., et al. (2008). ERα negative breast cancer cells restore response to endocrine therapy by combination treatment with both HDAC inhibitor and DNMT inhibitor. *Journal of Cancer Research and Clinical Oncology, 134*(8), 883–890.
29. Tellez, C. S., Grimes, M. J., Picchi, M. A., Liu, Y., March, T. H., Reed, M. D., et al. (2014). SGI-110 and entinostat therapy reduces lung tumor burden and reprograms the epigenome. *International Journal of Cancer Journal international du cancer, 135*(9), 2223–2231.
30. Mani, S. A., Guo, W., Liao, M.-J., Eaton, E. N., Ayyanan, A., Zhou, A. Y., et al. (2008). The epithelial-mesenchymal transition generates cells with properties of stem cells. *Cell, 133*(4), 704–715.
31. Zhang, Z. Y., Lu, Y. X., Zhang, Z. Y., Chang, Y. Y., Zheng, L., Yuan, L., et al. (2016). Loss of TINCR expression promotes proliferation, metastasis through activating EpCAM cleavage in colorectal cancer. *Oncotarget, 7*(16), 22639–22649.
32. Munz, M., Baeuerle, P. A., & Gires, O. (2009). The emerging role of EpCAM in cancer and stem cell signaling. *Cancer Research, 69*(14), 5627–5629.
33. Wang, Z. T., Chen, Z. J., Jiang, G. M., Wu, Y. M., Liu, T., Yi, Y. M., et al. (2016). Histone deacetylase inhibitors suppress mutant p53 transcription via HDAC8/YY1 signals in triple negative breast cancer cells. *Cellular Signalling, 28*(5), 506–515.
34. Stojanovic, N., Hassan, Z., Wirth, M., Wenzel, P., Beyer, M., Schäfer, C., et al. (2017). HDAC1 and HDAC2 integrate the expression of p53 mutants in pancreatic cancer. *Oncogene, 36*(13), 1804–1815.
35. Dumay, A., Feugeas, J. P., Wittmer, E., Lehmann-Che, J., Bertheau, P., Espie, M., et al. (2013). Distinct tumor protein p53 mutants in breast cancer subgroups. *International Journal of Cancer Journal International du Cancer, 132*(5), 1227–1231.
36. Dong, P., Karaayvaz, M., Jia, N., Kaneuchi, M., Hamada, J., Watari, H., et al. (2013). Mutant p53 gain-of-function induces epithelial-mesenchymal transition through modulation of the miR-130b-ZEB1 axis. *Oncogene, 32*(27), 3286–3295.
37. Jiang, F. Z., He, Y. Y., Wang, H. H., Zhang, H. L., Zhang, J., Yan, X. F., et al. (2015). Mutant p53 induces EZH2 expression and promotes epithelial-mesenchymal transition by disrupting p68-Drosha complex assembly and attenuating miR-26a processing. *Oncotarget, 6*(42), 44660–44674.
38. Piaskowski, S., Zawlik, I., Szybka, M., Kulczycka-Wojdala, D., Stoczynska-Fidelus, E., Bienkowski, M., et al. (2010). Detection of P53 mutations in different cancer types is improved by cDNA sequencing. *Oncology Letters, 1*(4), 717–721.
39. Holm, K., Grabau, D., Lovgren, K., Aradottir, S., Gruvberger-Saal, S., Howlin, J., et al. (2012). Global H3K27 trimethylation and EZH2 abundance in breast tumor subtypes. *Molecular Oncology, 6*(5), 494–506.
40. Hollestelle, A., Nagel, J. H., Smid, M., Lam, S., Elstrodt, F., Wasielewski, M., et al. (2010). Distinct gene mutation profiles among luminal-type and basal-type breast cancer cell lines. *Breast Cancer Research and Treatment, 121*(1), 53–64.
41. Tyagi, T., Treas, J. N., Mahalingaiah, P. K., & Singh, K. P. (2015). Potentiation of growth inhibition and epigenetic modulation by combination of green tea polyphenol and 5-aza-2′-deoxycytidine in human breast cancer cells. *Breast Cancer Research and Treatment, 149*(3), 655–668.

42. Dahn, M. L., Cruickshank, B. M., Jackson, A. J., Dean, C., Holloway, R. W., Hall, S. R., et al. (2020). Decitabine response in breast cancer requires efficient drug processing and is not limited by multidrug resistance. *Mol Cancer Ther, 19*(5), 1110–1122.
43. Srivastava, P., Paluch, B. E., Matsuzaki, J., James, S. R., Collamat-Lai, G., Karbach, J., et al. (2014). Immunomodulatory action of SGI-110, a hypomethylating agent, in acute myeloid leukemia cells and xenografts. *Leukemia Research, 38*(11), 1332–1341.
44. Srivastava, P., Paluch, B. E., Matsuzaki, J., James, S. R., Collamat-Lai, G., Taverna, P., et al. (2015). Immunomodulatory action of the DNA methyltransferase inhibitor SGI-110 in epithelial ovarian cancer cells and xenografts. *Epigenetics, 10*(3), 237–246.
45. Luker, A. J., Graham, L. J., Smith, T. M., Camarena, C., Zellner, M. P., Gilmer, J.-J. S., et al. (2020). The DNA methyltransferase inhibitor, guadecitabine, targets tumor-induced myelopoiesis and recovers T cell activity to slow tumor growth in combination with adoptive immunotherapy in a mouse model of breast cancer. *BMC Immunology, 21*(1), 1–15.
46. Fang, F., Munck, J., Tang, J., Taverna, P., Wang, Y., Miller, D. F., et al. (2014). The novel, small-molecule DNA methylation inhibitor SGI-110 as an ovarian cancer chemosensitizer. *Clinical Cancer Research: An Official Journal of the American Association for Cancer Research, 20*(24), 6504–6516.
47. Kuang, Y., El-Khoueiry, A., Taverna, P., Ljungman, M., & Neamati, N. (2015). Guadecitabine (SGI-110) priming sensitizes hepatocellular carcinoma cells to oxaliplatin. *Molecular Oncology, 9*(9), 1799–1814.
48. Yuan, Y. G., Peng, Q. L., & Gurunathan, S. (2017). Combination of palladium nanoparticles and tubastatin-A potentiates apoptosis in human breast cancer cells: A novel therapeutic approach for cancer. *International Journal of Nanomedicine, 12*, 6503–6520.
49. Huang, X., Gao, L., Wang, S., Lee, C. K., Ordentlich, P., & Liu, B. (2009). HDAC inhibitor SNDX-275 induces apoptosis in erbB2-overexpressing breast cancer cells via down-regulation of erbB3 expression. *Cancer Research, 69*(21), 8403–8411.
50. Srivastava, R. K., Kurzrock, R., & Shankar, S. (2010). MS-275 sensitizes TRAIL-resistant breast cancer cells, inhibits angiogenesis and metastasis, and reverses epithelial-mesenchymal transition in vivo. *Molecular Cancer Therapeutics, 9*(12), 3254–3266.
51. Shah, P., Gau, Y., & Sabnis, G. (2014). Histone deacetylase inhibitor entinostat reverses epithelial to mesenchymal transition of breast cancer cells by reversing the repression of E-cadherin. *Breast Cancer Research and Treatment, 143*(1), 99–111.
52. Schech, A., Kazi, A., Yu, S., Shah, P., & Sabnis, G. (2015). Histone deacetylase inhibitor entinostat inhibits tumor-initiating cells in triple-negative breast cancer cells. *Molecular Cancer Therapeutics, 14*(8), 1848–1857.
53. Lau, M. T., Klausen, C., & Leung, P. C. (2011). E-cadherin inhibits tumor cell growth by suppressing PI3K/Akt signaling via beta-catenin-Egr1-mediated PTEN expression. *Oncogene, 30*(24), 2753–2766.
54. Richon, V. M., Sandhoff, T. W., Rifkind, R. A., & Marks, P. A. (2000). Histone deacetylase inhibitor selectively induces p21WAF1 expression and gene-associated histone acetylation. *Proceedings of the National Academy of Sciences of the United States of America, 97*(18), 10014–10019.
55. De Carvalho Fischer, C., Hu, Y., Morreale, M., Lin, W. Y., Wali, A., Thakar, M., et al. (2018). Treatment with epigenetic agents profoundly inhibits tumor growth in leiomyosarcoma. *Oncotarget, 9*(27), 19379–19395.
56. Maetzel, D., Denzel, S., Mack, B., Canis, M., Went, P., Benk, M., et al. (2009). Nuclear signalling by tumour-associated antigen EpCAM. *Nature Cell Biology, 11*(2), 162–171.
57. Xu, J., Prosperi, J. R., Choudhury, N., Olopade, O. I., & Goss, K. H. (2015). β-Catenin is required for the tumorigenic behavior of triple-negative breast cancer cells. *PLoS One, 10*(2), e0117097.
58. De, P., Carlson, J. H., Wu, H., Marcus, A., Leyland-Jones, B., & Dey, N. (2016). Wnt-beta-catenin pathway signals metastasis-associated tumor cell phenotypes in triple negative breast cancers. *Oncotarget, 7*(28), 43124–43149.

59. Kim, T., Veronese, A., Pichiorri, F., Lee, T. J., Jeon, Y. J., Volinia, S., et al. (2011). p53 regulates epithelial-mesenchymal transition through microRNAs targeting ZEB1 and ZEB2. *The Journal of Experimental Medicine, 208*(5), 875–883.
60. Chang, C. J., Chao, C. H., Xia, W., Yang, J. Y., Xiong, Y., Li, C. W., et al. (2011). p53 regulates epithelial-mesenchymal transition and stem cell properties through modulating miRNAs. *Nature Cell Biology, 13*(3), 317–323.
61. Kleer, C. G., Cao, Q., Varambally, S., Shen, R., Ota, I., Tomlins, S. A., et al. (2003). EZH2 is a marker of aggressive breast cancer and promotes neoplastic transformation of breast epithelial cells. *Proceedings of the National Academy of Sciences of the United States of America, 100*(20), 11606–11611.
62. Pietersen, A. M., Horlings, H. M., Hauptmann, M., Langerod, A., Ajouaou, A., Cornelissen-Steijger, P., et al. (2008). EZH2 and BMI1 inversely correlate with prognosis and TP53 mutation in breast cancer. *Breast Cancer Research, 10*(6), R109.
63. Muller, P. A., & Vousden, K. H. (2013). p53 mutations in cancer. *Nature Cell Biology, 15*(1), 2–8.
64. Deb, G., Thakur, V. S., & Gupta, S. (2013). Multifaceted role of EZH2 in breast and prostate tumorigenesis: Epigenetics and beyond. *Epigenetics, 8*(5), 464–476.
65. Ramadoss, S., Chen, X., & Wang, C. Y. (2012). Histone demethylase KDM6B promotes epithelial-mesenchymal transition. *The Journal of Biological Chemistry, 287*(53), 44508–44517.
66. Zhang, J., Liu, D., Feng, Z., Mao, J., Zhang, C., Lu, Y., et al. (2016). MicroRNA-138 modulates metastasis and EMT in breast cancer cells by targeting vimentin. *Biomedicine & Pharmacotherapy = Biomedecine & pharmacotherapie, 77*, 135–141.
67. Yan, N., Xu, L., Wu, X., Zhang, L., Fei, X., Cao, Y., et al. (2017). GSKJ4, an H3K27me3 demethylase inhibitor, effectively suppresses the breast cancer stem cells. *Experimental Cell Research, 359*(2), 405–414.
68. Pathania, R., Ramachandran, S., Elangovan, S., Padia, R., Yang, P., Cinghu, S., et al. (2015). DNMT1 is essential for mammary and cancer stem cell maintenance and tumorigenesis. *Nature Communications, 6*, 6910.
69. Shin, E., Lee, Y., & Koo, J. S. (2016). Differential expression of the epigenetic methylation-related protein DNMT1 by breast cancer molecular subtype and stromal histology. *Journal of Translational Medicine, 14*, 87.
70. Cardenas, H., Vieth, E., Lee, J., Segar, M., Liu, Y., Nephew, K. P., et al. (2014). TGF-beta induces global changes in DNA methylation during the epithelial-to-mesenchymal transition in ovarian cancer cells. *Epigenetics, 9*(11), 1461–1472.
71. Russo, J., & Russo, I. H. (2014). *Techniques and methodological approaches in breast cancer research*. New York: Springer.
72. Santucci-Pereira, J., Su, Y., & Russo, J. (2018). *Proceedings of the American Association for Cancer Research Annual Meeting 2018*. Chicago/Philadelphia: AACR. Cancer Res 2018;78(13 Suppl): Abstract nr 3322.

Chapter 6
A New Treatment Strategy for *BRCA1*-Related Breast Cancer

Abstract Breast cancer susceptibility gene1 (*BRCA1*) is an important tumor suppressor gene that maintains genome integrity. *BRCA1*-associated breast cancer (BABC) is more likely to be triple-negative breast cancer (TNBC). In the past, BABC was treated in the same way for breast cancers without *BRCA1/2* mutations, which means no targeted therapy for those BABC with TNBC subtype. Our lab has been focusing on identifying new treatments for TNBC, and we have demonstrated that the combination of two epigenetic compounds, DNA methyltransferase inhibitor (DNMTi) SGI-110 (Guadecitabine) and histone deacetylation inhibitor (HDACi) MS275 (Entinostat), exerts a great anti-tumor activity both in vitro and in vivo against TNBC by simultaneously targeting multiple pathways. While screening the sensitivity of a panel of breast cancer cell lines to the treatment of SGI-110 and MS275, our data showed that the TNBC cell line with *BRCA1* mutation was more sensitive to these two compounds. We investigated more TNBC cell lines with *BRCA1* mutation, and here, we present our data that the proliferation of triple-negative BABC cells are sensitive to SGI, MS275, and the combination of SGI-110 with MS275 (SGI + MS). We furthermore demonstrated that in vivo treatment of SGI-110 and SGI + MS suppresses xenograft growth of *BRCA1* mutant breast cancer cell line Sum149pt in NOD SCID mice. Our study identifies an important role of SGI-110 and MS275 in suppressing the growth of BABC cells; we propose that the epigenetic compounds SGI-110 and MS275 could be used as a new treatment strategy for BABC.

Keywords Breast cancer susceptibility gene1 (*BRCA1*) · *BRCA1*-associated breast cancer (BABC) · Triple-negative breast cancer (TNBC) · DNA methyltransferase inhibitor (DNMTi). SGI-110 (Guadecitabine) · histone deacetylation inhibitor (HDACi) · MS275 (Entinostat)

6.1 Introduction

Breast cancer susceptibility gene1 (*BRCA1*) is an important tumor suppressor gene that maintains genome integrity [1]. Germline mutation of *BRCA1* leads to an increased risk of developing hereditary breast/ovarian cancer. The estimated cumulative risk of developing breast cancer by *BRCA*1 mutation is reported to be 46–87%

J. Russo, *The Future of Prevention and Treatment of Breast Cancer*,
https://doi.org/10.1007/978-3-030-72815-1_6

by age 70 years [2–4]. In addition, *BRCA1* carrier patients are significantly younger than sporadic breast cancer patients, and the *BRCA1*-assoicated breast cancer (BABC) is more likely to be triple-negative breast cancer (TNBC) [5, 6]. In the past, BABC was treated in the same way for breast cancers without *BRCA1/2* mutations, which means no targeted therapy for those BABC with TNBC subtype. The discovery that poly (ADP-ribose) polymerase (PARP) inhibitor (PARPi) could selectively cause DNA damage and synthetic lethality for cells with *BRCA1/2* mutations has been a breakthrough for the treatment of BABC [7]. Four PARPi have been approved by FDA for BABC treatment so far. Despite the success of this approach, like other targeted therapy, resistance to PARPi remains a clinical hurdle [8].

Our lab has been focusing on identifying new treatments for TNBC, and we have demonstrated that the combination of two epigenetic compounds, DNA methyltransferase inhibitor (DNMTi) SGI-110 (Guadecitabine) and histone deacetylation inhibitor (HDACi) MS275 (Entinostat), exerts a great anti-tumor activity both in vitro and in vivo against TNBC by simultaneously targeting multiple pathways [9]. While screening the sensitivity of a panel of breast cancer cell lines to the treatment of SGI-110 and MS275, our data showed that the TNBC cell line with *BRCA1* mutation was more sensitive to these two compounds. We investigated more TNBC cell lines with *BRCA1* mutation, and here, we present our data that the proliferation of triple-negative BABC cells are sensitive to SGI, MS275, and the combination of SGI-110 with MS275 (SGI + MS). We furthermore demonstrated that in vivo treatment of SGI-110 and SGI + MS suppresses xenograft growth of *BRCA1* mutant breast cancer cell line Sum149pt in NOD SCID mice. Our study identifies an important role of SGI-110 and MS275 in suppressing the growth of BABC cells; we propose that the epigenetic compounds SGI-110 and MS275 could be used as a new treatment strategy for BABC.

6.2 In Vitro Studies

In a previous study, we observed that the proliferation of TNBC cells were more sensitive to the treatment of SGI-110 and MS275 than luminal and HER2 type cells [9], and among the six TNBC cell lines being tested, Sum149pt was more sensitive than other TNBC cell lines [9]. Since Sum149pt is a cell line with *BRCA1* mutation, we evaluated three additional *BRCA1* mutant TNBC cell lines using MTT assay and found that with three consecutive days treatment, HCC1937 and MDAMB436 cell lines were also very sensitive to SGI-110 (Fig. 6.1a,b), presenting an IC50 lower than 500 nM. For MS275, we treated cells for 1 day only, and measured cell proliferation 4 days later. With this treatment schedule, only Sum149pt cell line was sensitive to MS275 when treated with doses lower than 4000 nM. (Figure 6.1a, b). However, when these cells were treated with the combination of SGI-110 with MS275 at low dose, cell proliferation was significantly inhibited compared to single agent alone (Fig. 6.1c, d), suggesting these two agents have a synergistic effect in inhibiting the growth of these BABC cells.

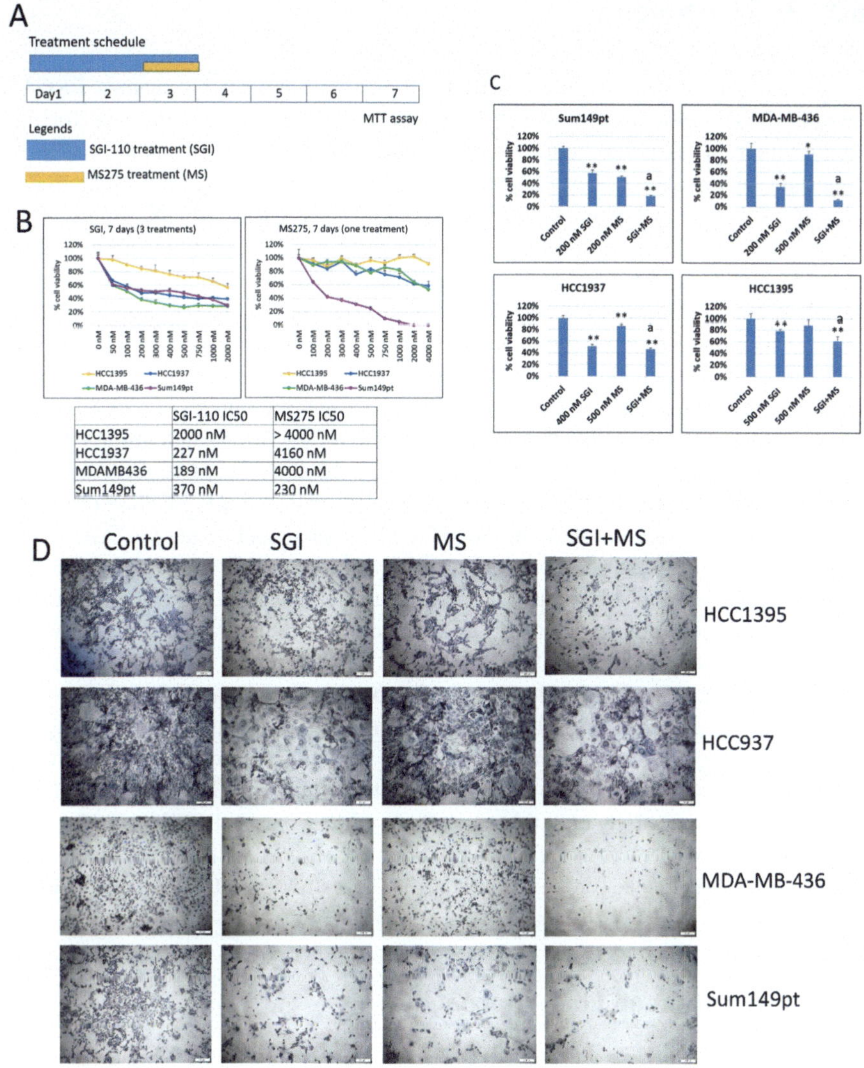

	SGI-110 IC50	MS275 IC50
HCC1395	2000 nM	> 4000 nM
HCC1937	227 nM	4160 nM
MDAMB436	189 nM	4000 nM
Sum149pt	370 nM	230 nM

Fig. 6.1 Effect of SGI-110 and MS275 on proliferation of TNBC cells with BRCA1 mutation. (**a**). Treatment schedule for MTT assay. Cells were plated in the 96-well plate, allowed to attach overnight, then treated with SGI-110 (SGI) for 3 days (daily medium change) or/and MS275 (MS) for 1 day, and cultured for additional 4 days; cell viability was measured by MTT assay at the end of 7-day culture. (**b**), Cell growth curves and IC50 measured by MTT assay. (**c**), Combined treatment of SGI with MS275 induces greater inhibition in cell proliferation. * indicates $p < 0.05$ when compared to control. ** indicates $p < 0.01$ when compared to control. a indicates $p < 0.01$ when compared to the combined treatment with single agent. (**d**), Phase contrast images of cells acquired after incubation with MTT for 4 hours. Doses used for SGI and MS for each cell line are indicated in (**c**). Scale bar, 200 μm. Magnification, 10× objective

6.3 In Vivo Studies

To evaluate the in vivo antitumor activity of SGI-110 and MS275 against *BRCA1* mutant TNBC cells, Sum149pt cell line and NOD SCID mice were used for the study. Sum149pt is a very aggressive cell line with high tumorigenicity and high metastatic potential. 100% of NOD SCID mice developed xenograft with an injection of 1,000,000 cells/mouse. After a xenograft tumor was palpable, mice were randomly divided into four groups, with eight mice per group, and the treatment of SGI-110 and MS275 was started. The treatment schedule selected (Fig. 6.2a) was based on the literatures [10] and our previous studies [9]. The wellbeing of the mouse was monitored throughout the treatment; no significant toxicity was observed with this treatment schedule; the body weight curves are shown in Fig. 6.2b. Tumors grew slowly in the first 4 weeks, but rapidly thereafter. The tumors of control group grew significantly faster than that of SGI-110 and SGI + MS treated groups during and after treatment (Fig. 6.3a). At sacrifice day the differences in tumor weight between control and SGI or SGI + MS group were very dramatic (median of tumor weight is 208 mg for control, 42 mg for SGI-110 treated, and 57 mg for SGI + MS) (Fig. 6.3b, c); the median tumor weight of SGI-110-treated group was reduced by 80% compared to control group. These data suggest that SGI-110 can exert a significant in vivo antitumor activity at a tolerable dose. Although we didn't observe an in vivo anti-tumor effect of MS275 in this study, it was reported by other researchers that MS275 suppresses Sum149pt xenograft growth in female athymic homozygous nu/nu mice when given by oral gavage daily for 8 weeks with a dose of 5 mg/kg [11], supporting our in vitro data that MS275 has an anti-tumor effect for triple-negative cancer with *BRCA1* mutation. Further investigation of SGI-110 in expanded BABC cell lines and study of anti-tumor effect of MS275 as well as the combination of SGI-110 with MS275 in different treatment schedules will be very useful for providing more evidence for the potential use of this strategy to treat BABC patients.

6.4 Innovation of This Approach

The treatment of TNBC is a clinical challenge, in the subset of *BRCA1*-associated TNBC; although PARPi has been used for metastatic breast cancer treatment in addition to the traditional chemotherapy such as Taxanes and Platinum agents, a recent clinical trial revealed the PARPi Talazoparib didn't significantly improve overall survival over chemotherapy in patients with germline BRCA1/2-mutated HER2-negative advanced breast cancer [12]. In addition, resistance to these treatments is still a serious clinical problem [13]. These shortcomings have led to search for more effective treatments for this disease. The concept of synthetic lethality that cell death would be efficiently induced by simultaneous loss of function of several key molecules led to the exploration of treatments with various combination [14, 15]. Clinical trials in breast cancer patients with *BRCA* mutation using PARPi

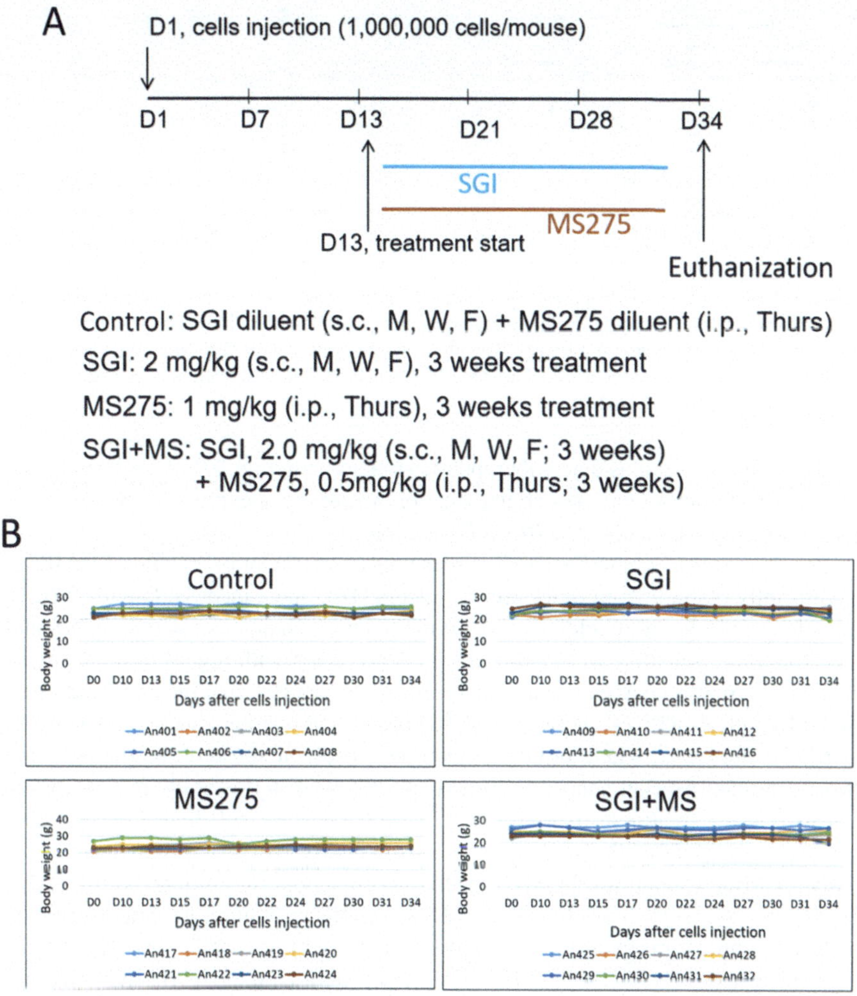

Fig. 6.2 In vivo treatment schedule and body weight of animals in different groups. (**a**), Treatment schedule. Female NOD SCID mice at 8 weeks old were used for the study. 1,000,000 Sum149pt cells suspended in 0.1 ml of the mixture of the sterile phosphate buffered saline (PBS) and Matrigel (1:1 mixture) were injected subcutaneously to the mouse mammary fat pad of the abdominal region. Treatment was started 13 days post cells injection when a xenograft was palpable and stopped on 31 days after 3 weeks of treatment. Mice were euthanized 34 days post cells injection when the largest tumor reached the diameter of 15 mm. (**b**), Body weight of animals from the day of cells injection to the day of euthanization. Each line represents one animal

combined with other treatments including paclitaxel, cisplatin, and compounds of immunotherapy are ongoing. We demonstrated that treatment of TNBC cells with two epigenetic compounds, SGI-110 and MS275, or their combination exerts a great antitumor activity [9], especially for the TNBC with BRCA1 mutation. These findings led us to hypothesize that the epigenetic compounds such as SGI-110 and

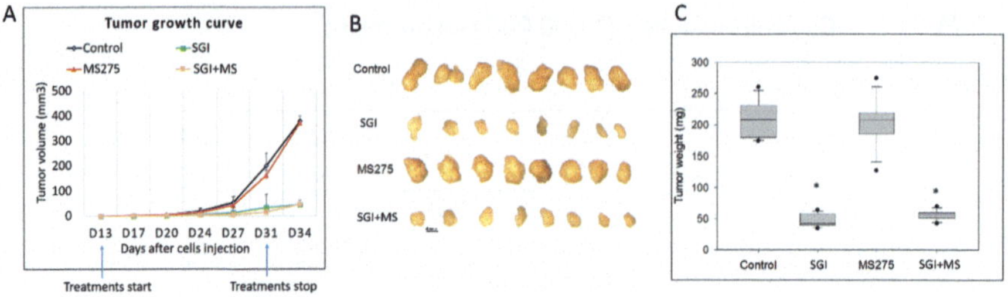

Fig. 6.3 SGI and SGI+MS suppress Sum149pt xenograft tumor growth at a tolerable dose. (**a**), Tumor growth curve calculated by the formula: volume = 1/2*length*width^2. (**b**), Picture of xenografts at sacrifice. Scale bar: 4 mm. (**c**), Box plots of xenograft weight. The line in the box shows the median value. * indicates $p < 0.05$ by One Way Analysis followed with all pairwise multiple comparison procedures with Turkey test

MS275 could be used as a new therapeutic strategy targeting *BRCA1* related breast cancer. This hypothesis is conceptually novel by seeking to explore a new treatment for breast cancer patients with *BRCA1* mutation.

Breast cancer is a heterogeneous disease, DNA-methylation profile of BABC is significantly distinct from other breast cancers [16], and cancer cells with *BRCA1* mutation might be more addictive to the methylation changes of certain genes and therefore more sensitive to DNMTi. There is no report with regard to the using of SGI-110 in treating BABC so far. One recent study reported that the combination of SGI-110 with PARPi Talazoparib inhibited breast and ovarian cancers independent of *BRCA* mutations; when looking at the data closely, it showed that the only one *BRCA1* mutant ovarian cancer cell line COV362 used in this study is much more sensitive to the treatment than all other *BRCA1* wild-type cell lines [17]. Another study also showed that the progenitor cells from breast tissues of tumor-free *BRCA1* mutant carrier women are highly sensitive to epigenetic inhibition; both DNMTi Decitabine and HDACi Trichostatin A (TSA) significantly inhibit colony formation of *BRCA1* mutant progenitor cells but not *BRCA1* wild type cells [18]. These findings support our data that SGI-110 has an anti-tumor effect in BABC cells and might be partially through targeting cancer stem cells. MS275 has been studied as a treatment for advanced breast cancer and other types of cancer and has been shown to inhibit breast cancer cells by inducing acetylated histone H1, reprograming tumor infiltrating myeloid-derived suppressor cells, inhibiting tumor-initiating cells, suppressing the epithelial-mesenchymal transition and DNA repair protein expression, downregulating pro-angiogenesis genes, and re-expressing anti-angiogenic genes and tumor suppressor genes [19–24]. In addition, it is accepted that both genetic and epigenetic alterations play important roles in tumor progression. Besides the *BRCA1* mutation, BABC often have p53 mutation; we have shown that the combination of SGI-110 with MS275 decreases the expression of mutant p53 in TNBC cells. Taken together, the combination of SGI-110 with MS275 might exert superior efficacy against BABC by targeting multiple pathways. The recognition of BABC as a specific target for this new treatment strategy could broaden the treatment option and finally benefit these patients.

6.5 Clinical Impact

TNBC is more common in patients with *BRCA1* mutation than those of non-*BRCA*-carriers [25]; it is reported up to 66% of the breast cancers in *BRCA1* carriers are TNBC [26]. TNBC is considered to be more aggressive and has a poorer prognosis than other breast cancers. Epigenetic therapeutic has been a promising therapy for cancers based on epigenetic abnormalities. Both SGI-110 and MS275 are in clinical trials for treatment of various cancers. SGI-110 has been used in the trials for leukemia, metastatic melanoma, ovarian cancer, hepatocellular carcinoma, etc. MS275 is currently undergoing clinical trials in patients with advanced breast cancer and other cancers. The result presented here has a translational significance that will provide evidence for using these two compounds in the clinical trial for BABC as a tailored therapy, enabling further advances toward personalized medical care. In addition, the mutation in *BRCA1* also increases the risk of developing ovarian cancer; the treatment strategy we proposed here might also apply to the ovarian cancer patient with *BRCA1* mutation.

6.6 Industry Relationships

The main chemical, SGI-110 (Guadecitabine), used in this proposal is provided by the Astex Pharmaceuticals Inc. under a contract between Astex Pharmaceuticals Inc. and Fox Chase Cancer Center.

References

1. Venkitaraman, A. R. (2002). Cancer susceptibility and the functions of BRCA1 and BRCA2. *Cell, 108*(2), 171–182.
2. Ford, D., Easton, D. F., Bishop, D. T., Narod, S. A., & Goldgar, D. E. (1994). Risks of cancer in BRCA1-mutation carriers. Breast Cancer Linkage Consortium. *Lancet, 343*(8899), 692–695.
3. Antoniou, A., Pharoah, P. D., Narod, S., Risch, H. A., Eyfjord, J. E., Hopper, J. L., et al. (2003). Average risks of breast and ovarian cancer associated with BRCA1 or BRCA2 mutations detected in case Series unselected for family history: A combined analysis of 22 studies. *American Journal of Human Genetics, 72*(5), 1117–1130.
4. Kuchenbaecker, K. B., Hopper, J. L., Barnes, D. R., Phillips, K. A., Mooij, T. M., Roos-Blom, M. J., et al. (2017). Risks of breast, ovarian, and contralateral breast cancer for BRCA1 and BRCA2 mutation carriers. *Journal of the American Medical Association, 317*(23), 2402–2416.
5. Gonzalez-Angulo, A. M., Timms, K. M., Liu, S., Chen, H., Litton, J. K., Potter, J., et al. (2011). Incidence and outcome of BRCA mutations in unselected patients with triple receptor-negative breast cancer. *Clinical Cancer Research, 17*(5), 1082–1089.
6. Arpino, G., Pensabene, M., Condello, C., Ruocco, R., Cerillo, I., Lauria, R., et al. (2016). Tumor characteristics and prognosis in familial breast cancer. *BMC Cancer, 16*(1), 1–8.
7. McCann, K. E., & Hurvitz, S. A. (2018). Advances in the use of PARP inhibitor therapy for breast cancer. *Drugs in Context, 7*, 212540.

8. Gogola, E., Rottenberg, S., & Jonkers, J. (2019). Resistance to PARP inhibitors: Lessons from preclinical models of BRCA-associated cancer. *Annual Review Cancer Biology, 3*, 235–254.
9. Su, Y., Hopfinger, N. R., Nguyen, T. D., Pogash, T. J., Santucci-Pereira, J., & Russo, J. (2018). Epigenetic reprogramming of epithelial mesenchymal transition in triple negative breast cancer cells with DNA methyltransferase and histone deacetylase inhibitors. *Journal of Experimental & Clinical Cancer Research, 37*(1), 314.
10. Tellez, C. S., Grimes, M. J., Picchi, M. A., Liu, Y., March, T. H., Reed, M. D., et al. (2014). SGI-110 and entinostat therapy reduces lung tumor burden and reprograms the epigenome. *International Journal of Cancer, 135*(9), 2223–2231.
11. Torres-Adorno, A. M., Lee, J., Kogawa, T., Ordentlich, P., Tripathy, D., Lim, B., et al. (2017). Histone deacetylase inhibitor enhances the efficacy of MEK inhibitor through NOXA-mediated MCL1 degradation in triple-negative and inflammatory breast cancer. *Clinical Cancer Research, 23*(16), 4780–4792.
12. Litton, J., Hurvitz, S., Mina, L., Rugo, H., Lee, K.-H., Gonçalves, A., et al. (2020). Talazoparib versus chemotherapy in patients with germline BRCA1/2-mutated HER2-negative advanced breast cancer: Final overall survival results from the EMBRACA trial. *Annals of Oncology, 31*, 1526–1535.
13. Rose, M., Burgess, J. T., O'Byrne, K., Richard, D. J., & EJFiC, B. (2020). PARP inhibitors: Clinical relevance, mechanisms of action and tumor resistance. *Frontiers in Cell and Development Biology, 8*, 564601.
14. Shaheen, M., Allen, C., Nickoloff, J. A., & Hromas, R. (2011). Synthetic lethality: Exploiting the addiction of cancer to DNA repair. *Blood, 117*(23), 6074–6082.
15. Dedes, K. J., Wilkerson, P. M., Wetterskog, D., Weigelt, B., Ashworth, A., & Reis-Filho, J. S. (2011). Synthetic lethality of PARP inhibition in cancers lacking BRCA1 and BRCA2 mutations. *Cell Cycle, 10*(8), 1192–1199.
16. Flanagan, J. M., Cocciardi, S., Waddell, N., Johnstone, C. N., Marsh, A., Henderson, S., et al. (2010). DNA methylome of familial breast cancer identifies distinct profiles defined by mutation status. *American Journal of Human Genetics, 86*(3), 420–433.
17. Pulliam, N., Fang, F., Ozes, A. R., Tang, J., Adewuyi, A., Keer, H., et al. (2018). An effective epigenetic-PARP inhibitor combination therapy for breast and ovarian cancers independent of BRCA mutations. *Clinical Cancer Research, 24*(13), 3163–3175.
18. Casey, A. E., Sinha, A., Singhania, R., Livingstone, J., Waterhouse, P., Tharmapalan, P., et al. (2018). Mammary molecular portraits reveal lineage-specific features and progenitor cell vulnerabilities. *The Journal of Cell Biology, 217*(8), 2951–2974.
19. Androutsopoulos, V. P., & Spandidos, D. A. (2017). Antiproliferative effects of TSA, PXD101 and MS275 in A2780 and MCF7 cells: Acetylated histone H4 and acetylated tubulin as markers for HDACi potency and selectivity. *Oncology Reports, 38*(6), 3412–3418.
20. Christmas, B. J., Rafie, C. I., Hopkins, A. C., Scott, B. A., Ma, H. S., Cruz, K. A., et al. (2018). Entinostat converts immune-resistant breast and pancreatic cancers into checkpoint-responsive tumors by reprogramming tumor-infiltrating MDSCs. *Cancer Immunology Research, 6*(12), 1561–1577.
21. Merino, V. F., Nguyen, N., Jin, K., Sadik, H., Cho, S., Korangath, P., et al. (2016). Combined treatment with epigenetic, differentiating, and chemotherapeutic agents cooperatively targets tumor-initiating cells in triple-negative breast cancer. *Cancer Research, 76*(7), 2013–2024.
22. Schech, A., Kazi, A., Yu, S., Shah, P., & Sabnis, G. (2015). Histone deacetylase inhibitor entinostat inhibits tumor-initiating cells in triple-negative breast cancer cells. *Molecular Cancer Therapeutics, 14*(8), 1848–1857.
23. Kiweler, N., Wünsch, D., Wirth, M., Mahendrarajah, N., Schneider, G., Stauber, R. H., et al. (2020). Histone deacetylase inhibitors dysregulate DNA repair proteins and antagonize metastasis-associated processes. *Journal of Cancer Research and Clinical Oncology, 146*(2), 343–356.
24. Maiti, A., Qi, Q., Peng, X., Yan, L., & Takabe, K. (2019). Hait NCJIjoo. Class I histone deacetylase inhibitor suppresses vasculogenic mimicry by enhancing the expression of tumor sup-

pressor and anti-angiogenesis genes in aggressive human TNBC cells. *International Journal of Oncology, 55*(1), 116–130.
25. Chen, H., Wu, J., Zhang, Z., Tang, Y., Li, X., Liu, S., et al. (2018). Association between BRCA status and triple-negative breast Cancer: A meta-analysis. *Frontiers in Pharmacology, 9*, 909.
26. Peshkin, B. N., Alabek, M. L., & Isaacs, C. (2010). BRCA1/2 mutations and triple negative breast cancers. *Breast Disease, 32*(1–2), 25–33.

Chapter 7
Present Options in the Prevention of Breast Cancer

Abstract Among the risk factors for the development of breast cancers are age, reproductive factors and sex hormone exposure, biopsy history and results, and family history of breast and ovarian cancer. Several models have been developed for predicting risk, and among them are models for a high-risk population, especially those with precancerous changes, including the Breast Cancer Risk Assessment Tool (BCRAT or GAIL Model) or the International Breast Cancer Intervention Study (IBIS) or Tyrer-Cuzick model and those more specific to the probability of carrying a BRCA1 or BRCA2 mutation including the BOADICEA, CLAUS, and BRCAPRO (an upcoming enhancement to the Breast and Ovarian Analysis of Disease Incidence and Carrier Estimation Algorithm (BOADICEA) will include PALB2 and RAD51). Mammographic Breast Density has emerged as a reliable way to determine risk, for example, having dense breast tissue by mammogram is linked to an increased risk of developing breast cancer compared with women with less dense breast tissue. The development of better mammographic techniques and software like the VOLPARA that objectively measures breast density has added a new dimension in the assessment of risk. Many lifestyle factors, including physical inactivity, poor diet, alcohol intake, and high body mass index (BMI), have been associated with an increased risk of breast cancer and mortality. In summary among the cancer prevention guidelines available, there is a general consensus that maintaining a healthy weight, undertaking at least 150 minutes of moderate-intensity exercise per week, limiting alcohol consumption, and eating a plant-based diet are beneficial. Although weight control throughout life appears to prevent cases after menopause, the adherence to a healthy dietary pattern does not have specific effects on breast cancer risk but remains important as it reduces the risk for other common diseases, such as cardiovascular disease (CVD), diabetes, and dementia. Interesting smoking during adolescence or early adulthood increases later risk of breast cancer. Early first full-term pregnancy (FFTP) that occurs before the age of 20–25 has been shown to reduce a woman's lifetime breast cancer risk by as much as 50%, and risk is further reduced by another 7% for each subsequent pregnancy. Thus, an early FFTP is widely known as a protective factor against breast cancer. The first documented observation of this association was by Bernardino Ramazzini in 1700. Since then, many other data in concordance with this phenomenon have been published, showing that a FFTP before the age of 20 or 25 reduces lifetime risk by at least 38%. In contrast, women who have had their FFTP after the age of 30 show an increase

J. Russo, *The Future of Prevention and Treatment of Breast Cancer*,
https://doi.org/10.1007/978-3-030-72815-1_7

breast cancer risk. The importance of reproductive factors in breast cancer prevention has been emphasized in the literature.

Keywords Risk factors · Breast Cancer Risk Assessment Tool (BCRAT or GAIL Model) · the International Breast Cancer Intervention Study (IBIS) or Tyrer-Cuzick model · BOADICEA · CLAUS · BRCAPRO · Mammographic Breast Density lifestyle factors · First full-term pregnancy (FFTP)

7.1 Introduction

When the term breast cancer prevention is used, it means the "action taken to decrease the chance of developing cancer" [1–3]. This requires avoiding risk factors such as smoking, obesity, lack of exercise, alcohol consumption, and radiation exposure, and increasing protective factors such as engaging in regular physical activity, maintaining a healthy weight, a healthy diet [1], and understanding of the reproductive factors like early first full-term pregnancy [4].

Because the human nature of avoid effort, other alternative strategies have been suggested such as chemoprevention and risk-reducing surgery. Therefore, it is clear that there is not a perfect guideline that prevent breast cancer creating the need to identify a target population by determining the risk for a woman to develop the disease.

Among the risk factors are age, reproductive factors and sex hormone exposure, biopsy history and results, and family history of breast and ovarian cancer. Several models have been developed for predicting risk, and among them are models for a high-risk population, especially those with precancerous changes, including the Breast Cancer Risk Assessment Tool (BCRAT or GAIL Model) [5, 6] or the International Breast Cancer Intervention Study (IBIS) or Tyrer-Cuzick model [7] and those more specific to the probability of carrying a BRCA1 or BRCA2 mutation including the BOADICEA, CLAUS, and BRCAPRO (an upcoming enhancement to the Breast and Ovarian Analysis of Disease Incidence and Carrier Estimation Algorithm (BOADICEA) will include PALB2 and RAD51) [8–13]. Many other authoritative publications in developing model of risk have been discussed in the literature [14–17] highlighting the interest and need of better selective methods for detecting the population at risk.

Mammographic Breast Density has emerged as a reliable way to determine risk, for example, having dense breast tissue by mammogram is linked to an increased risk of developing breast cancer compared with women with less dense breast tissue [18]. The development of better mammographic techniques and software like the VOLPARA that objectively measures breast density has added a new dimension in the assessment of risk [19]. Although the diagnosis of intraepithelial neoplasia is associated with risk mainly when detected by mammography, the concept to measure that as a preventive strategy is already too late for being considered in the management of real primary breast cancer prevention [20, 21].

7.2 Lifestyle Related to Breast Cancer Prevention

Many lifestyle factors, including physical inactivity, poor diet, alcohol intake, and high body mass index (BMI), have been associated with an increased risk of breast cancer and mortality [22–31]. Based on these studies numerous organizations, including the American Cancer Society (ACS), have produced cancer prevention guidelines [30, 32]. Adherence to four lifestyle factors such as healthy body weight, physical activity, diet, and alcohol intake has shown 23% decreased risk of all-cause mortality comparing high to low adherence quintiles [33].

In a recent study [34] it was found that four of the six individual modifiable risk factors, BMI, physical activity, alcohol intake, and tobacco use, were significantly related to breast cancer incidence. There were, however, no individual risk factors significantly associated with breast cancer-specific mortality, and the models did not show a significant association between lifestyle factors and breast cancer-specific mortality. There is consistency in the data reported in the literature which found between a 16% and 60% reduction in breast cancer incidence for the highest level of prevention guidelines adherence compared to the lowest level, primarily among postmenopausal women [35–39]. Daily TV watching may also allow for simultaneous unhealthy lifestyle habits, such as poor dietary choices and increased smoking or alcohol consumption, and is therefore an important and modifiable facet of sedentary behavior [40, 41]. All of these life factors need to be also taken with caution because the majority of the data are obtained by self-reporting using questionnaire-based collection.

In summary among the cancer prevention guidelines available, there is a general consensus that maintaining a healthy weight, undertaking at least 150 minutes of moderate-intensity exercise per week, limiting alcohol consumption, and eating a plant-based diet [42] are beneficial. Although weight control throughout life appears to prevent cases after menopause, the adherence to a healthy dietary pattern does not have specific effects on breast cancer risk but remains important as it reduces the risk for other common diseases, such as cardiovascular disease (CVD), diabetes, and dementia. Interesting smoking during adolescence or early adulthood increases later risk of breast cancer [42].

The message derived from these studies is that lifestyle factors modify risk among high-risk women with a family history and those with typical risk of the general population. The mechanism of action by which these life factors influence risk is not clearly understood; however, intuitively education and changes in lifestyle must be implemented since early age when the rapidly developing breast is particularly susceptible to carcinogenesis. Although we do not know all the facts, there is experiment evidence that the window of susceptibility of exposure to environmental carcinogens is important in the susceptibility to develop breast cancer, and therefore looking at early age is probably the most important lesson that we have learned and that we must implement in the future of breast cancer prevention [43–44].

7.3 Hormone Prevention

The use of aspirin, statins, and anti-hypertensive agents also called chemo-preventive drugs had a major impact on the incidence of cardiovascular disease [45]. The endocrine control of the breast and the understanding on the role of estrogen in breast cancer [46] has been a quantum leap for developing agents that specifically modulate the estrogen receptors [47–53] or inhibit the activation of estrogen like it is observed using aromatase inhibitors.

Tamoxifen and raloxifene are selective estrogen receptor modulators (SERMS) that reduced the risk of breast cancer [54], and both were similar in reducing developing a new invasive breast cancer by almost 50%. Although tamoxifen had a significant reduction in the incidence of DCIS (ductal carcinoma in situ) compared with raloxifene, it had unwanted side effects such as uterine cancer and thromboembolic events [55, 56]. After 20 years of use the effect of tamoxifen reduced the rate of breast cancer by about 30% for ER-positive breast cancer, but had no effect for ER-negative breast cancer. Although this benefit still tamoxifen does not show a survival advantage [57].

The use of aromatase inhibitor exemestane tested in the Mammary Prevention 3 (MAP.3) resulted in a 65% reduction of ER-positive invasive breast cancers [58]. Anastrozole, another aromatase inhibitor, was tested in the International Breast Cancer Intervention Study II (IBIS-II) showing a 53% reduction of invasive breast cancer and DCIS and a 58% reduction in ER-positive invasive breast cancer [59].

The combined use of exemestane plus ovarian function suppression, meaning ovariectomy, was superior to tamoxifen alone or tamoxifen plus ovarian function suppression in pre-menopausal women with hormone receptor-positive early breast cancer. Treatment with exemestane plus ovarian function suppression reduced the risk of any second primary invasive cancer by 28%, and reduced the risk of invasive breast cancer recurrence by 34%, compared with treatment with tamoxifen plus ovarian function suppression [60–62].

Data from 9 major trials involving more than 83,000 participants shows that selective estrogen receptor modulators (SERMs) reduce breast cancer incidence by 38% [45]. Combined results from 2 large trials with 8424 participants show that aromatase inhibitors (AIs) reduce breast cancer incidence by 53% [45].

An important warning is that although these data are encouraging, the use of chemo-preventive compounds like selective estrogen receptor modulators, aromatase indictors, and ovarian suppression *must not be considered preventive strategies but interventions for breast cancer reduction.*

7.4 Reproductive Factor and Breast Cancer Prevention

Early first full-term pregnancy (FFTP) that occurs before the age of 20–25 has been shown to reduce a woman's lifetime breast cancer risk by as much as 50%, and risk is further reduced by another 7% for each subsequent pregnancy [63]. Thus, an early

FFTP is widely known as a protective factor against breast cancer. The first documented observation of this association was by Bernardino Ramazzini in 1700 [64]. Since then, many other data in concordance with this phenomenon have been published, showing that a FFTP before the age of 20 or 25 reduces lifetime risk by at least 38% [64]. In contrast, women who have had their FFTP after the age of 30 show an increased breast cancer risk [63]. The importance of reproductive factors in breast cancer prevention has been emphasized in the literature [65–70] and will be the subject of three chapters in this book.

7.5 Clinical Trials in Breast Cancer Prevention

In the study of Arum et al. [71], it has shown that seventy percent of breast cancers are estrogen receptor (ER)-positive, and potentially treatable or preventable by anti-estrogenic agents. Therefore, the large, phase III randomized controlled breast cancer prevention trials have tested and are testing only hormonal drugs designed to antagonize the carcinogenic effect of endogenous estrogen; these agents are either selective estrogen receptor modulators (SERMs) or aromatase inhibitors (AIs). A number of phase I and II trials using tissue-derived surrogate endpoint biomarkers (SEBs) as outcomes have been implemented. These smaller trials address prevention not only of ER-negative but also ER-positive breast cancers, since approximately 50% of the latter have been shown to be resistant to the estrogen-targeting drugs used in the large trials. Issues of importance in these smaller trials include choice of agent, selection of appropriate trial participants, trial design, method of access to breast tissue in women without cancer, selection and monitoring of SEBs, and monitoring of drug toxicity [71]. Again these are not really primary breast cancer prevention trials but are only reduction strategies for women already diagnosed with breast cancer.

7.6 Other Chemo-Preventive Strategies Tested in Breast Cancer

Metformin was first introduced as a treatment for non-insulin-dependent (type 2) diabetes in the late 1950s and is currently a target for both primary and secondary breast cancer prevention, as well as other cancers (including colorectal, prostate, and endometrial). Much of the human data on metformin and cancer has come from epidemiologic studies of people with diabetes and retrospective studies [72–77], and there are not definitive trials showing benefits of this drug in primary prevention of breast cancer.

There are studies, both pre-clinical and clinical, that are evaluating whether vitamin D3 supplementation can alter breast density and other markers of breast cancer risk [78], but to date it has not been shown to have any effect.

There are also no clinical data reported to date to provide a strong association between statin use and breast cancer risk in average-risk women, although several observational cohort studies have demonstrated reduced risk of breast cancer recurrence in women using statins [79].

It has been shown that increased intake of omega-3 fatty acids (n−3) associated with decreased omega-6 (n−6), resulting in a higher n−3/n−6 ratio compared with the Western diet, is inversely associated with breast cancer risk [80]. The studies have shown some data demonstrating adherence and safety, changes in breast adiposity and density, and modulation of precancerous breast disease [81] but not evidence that can be used as a primary preventive modality [82].

The use of prophylactic mastectomy (PM) is generally reserved for those at hereditary cancer risk [83–86]. PM has been shown to reduce the risk of breast cancer in high-risk women by more than 90% [87], and studies have demonstrated that PM specifically reduces the risk of breast cancer in women with mutations in BRCA1 or BRCA2 by 89.5– 100% [88–95].

Prophylactic bilateral salpingo-oophorectomy has also been shown to reduce the risk of breast cancer by up to 68% in women who have not been diagnosed with breast cancer with BRCA1 and BRCA2 mutations [96–98]. The risk reduction of a second primary breast cancer with prophylactic oophorectomy is conflicting [99–101], and the sad part of the present situation is that there is no alternative or preventive strategy for these women.

References

1. National Cancer Institute. Breast cancer prevention PDQ®. http://www.cancer.gov/cancertopics/pdq/prevention/breast/Patient/page1.
2. Klemp, J. R. (2015). Breast cancer prevention across the cancer care continuum. *Seminars in Oncology Nursing, 31*, 89–99.
3. Colditz, G. A., & Bohlke, K. (2014). Priorities for the primary prevention of breast cancer. *CA. A Cancer Journal for Clinicians, 64*, 186–194.
4. Russo, J., Tay, L. K., & Russo, I. H. (1982). Differentiation of the mammary gland and susceptibility to carcinogenesis. *Breast Cancer Research and Treatment, 2*, 5–73.
5. Gail, M. H., Brinton, L. A., Byar, D. P., et al. (1989). Projecting individualized probabilities of developing breast cancer for white females who are being examined annually. *Journal of the National Cancer Institute, 81*, 1879–1886.
6. Costantino, J. P., Gail, M. H., Pee, D., et al. (1999). Validation studies for models projecting the risk of invasive and total breast cancer incidence. *Journal of the National Cancer Institute, 91*, 1541–1548.
7. Tyrer, J., Duffy, S. W., & Cuzick, J. (2004). A breast cancer prediction model incorporating familial and personal risk factors. *Statistics in Medicine, 23*, 1111–1130.
8. Antoniou, A. C., Pharoah, P. P., Smith, P., & Easton, D. F. (2004). The BOADICEA model of genetic susceptibility to breast and ovarian cancer. *British Journal of Cancer, 91*, 1580–1590.

9. Antoniou, A., Pharoah, P. D., Narod, S., et al. (2003). Average risks of breast and ovarian cancer associated with BRCA1 or BRCA2 mutations detected in case series unselected for family history: A combined analysis of 22 studies. *American Journal of Human Genetics, 72*, 1117–1130.
10. Parmigiani, G., Berry, D., & Aguilar, O. (1998). Determining carrier probabilities for breast cancer-susceptibility genes BRCA1 and BRCA2. *American Journal of Human Genetics, 62*, 145–158.
11. Claus, E. B., Risch, N., & Thompson, W. D. (1994). Autosomal dominant inheritance of early onset breast cancer. *Cancer, 73*, 643–651.
12. Berry, D. A., Iversen, E. S., Jr., Gudbjartsson, D. F., et al. (2002). BRCAPRO validation, sensitivity of genetic testing of BRCA1/BRCA2, and prevalence of other breast cancer susceptibility genes. *Journal of Clinical Oncology, 20*, 2701–2712.
13. Apostolou, P., & Fostira, F. (2013). Hereditary breast cancer: the era of new susceptibility genes. *Biotechnology Research International, 2013*, 747318.
14. Gail, M. H., & Pfeiffer, R. M. (2018). breast cancer risk model requirements for counseling, prevention, and screening. *Journal of the National Cancer Institute, 110*, 994–1002.
15. Sun, Y. S., Zhao, Z., Yang, Z. N., Xu, F., Lu, H. J., Zhu, Z. Y., Shi, W., Jiang, J., Yao, P. P., & Zhu, H. P. (2017). Risk factors and preventions of breast cancer. *International Journal of Biological Sciences, 13*, 1387–1397.
16. Winn, A. N., Ekwueme, D. U., Guy, G. P., Jr., & Neumann, P. J. (2016). Cost-utility analysis of cancer prevention, treatment, and control: a systematic review. *American Journal of Preventive Medicine, 50*, 241–248.
17. Howell, A., Anderson, A. S., Clarke, R. B., Duffy, S. W., Evans, D. G., Garcia-Closas, M., Gescher, A. J., Key, T. J., Saxton, J. M., & Harvie, M. N. (2014). Risk determination and prevention of breast cancer. *Breast Cancer Research, 16*, 446.
18. Boyd, N. F., Martin, L. J., Yaffe, M. J., & Minkin, S. (2011). Mammographic density and breast cancer risk: Current understanding and future prospects. *Breast Cancer Research, 13*, 223.
19. Eng, A., Gallant, Z., Shepherd, J., et al. (2014). Digital mammographic density and breast cancer risk: A case-control study of six alternative density assessment methods. *Breast Cancer Research, 16*, 439.
20. Hartmann, L. C., Degnim, A. C., Santen, R. J., Dupont, W. D., & Ghosh, K. (2015). Atypical hyperplasia of the breast-risk assessment and management options. *The New England Journal of Medicine, 372*, 78–89.
21. Hartmann, L. C., Radisky, D. C., Frost, M. H., et al. (2014). Understanding the premalignant potential of atypical hyperplasia through its natural history: A longitudinal cohort study. *Cancer Prevention Research, 7*, 211–217.
22. Ibrahim, E. M., & Al-Homaidh, A. (2011). Physical activity and survival after breast cancer diagnosis: Meta-analysis of published studies. *Medical Oncology, 28*, 753–765.
23. Berube, S., Lemieux, J., Moore, L., Maunsell, E., & Brisson, J. (2014). Smoking at time of diagnosis and breast cancer-specific survival: new findings and systematic review with meta-analysis. *Breast Cancer Research, 16*, R42.
24. Chan, D. S., Vieira, A. R., Aune, D., Bandera, E. V., Greenwood, D. C., McTiernan, A., et al. (2014). Body mass index and survival in women with breast cancer-systematic literature review and meta-analysis of 82 follow-up studies. *Annals of Oncology, 25*, 1901–1914.
25. Kroenke, C.H., Fung, T.T., Hu, F.B., . Holmes, M.D. Dietary patterns and survival after breast cancer diagnosis. Journal of Clinical Oncology 2005, 23:9295–9303.
26. Lew, J. Q., Freedman, N. D., Leitzmann, M. F., Brinton, L. A., Hoover, R. N., Hollenbeck, A. R., et al. (2009). Alcohol and risk of breast cancer by histologic type and hormone receptor status in postmenopausal women: The NIH-AARP Diet and Health Study. *American Journal of Epidemiology, 170*, 308–317.
27. McKenzie, F., Ferrari, P., Freisling, H., Chajes, V., Rinaldi, S., de Batlle, J., et al. (2015). Healthy lifestyle and risk of breast cancer among postmenopausal women in the European

Prospective Investigation into Cancer and Nutrition cohort study. *International Journal of Cancer, 136*, 2640–2648.
28. Nichols, H. B., Trentham-Dietz, A., Egan, K. M., Titus-Ernstoff, L., Holmes, M. D., Bersch, A. J., et al. (2009). Body mass index before and after breast cancer diagnosis: Associations with all-cause, breast cancer, and cardiovascular disease mortality. *Cancer Epidemiology, Biomarkers & Prevention, 18*, 1403–1409.
29. Schmid, D., & Leitzmann, M. F. (2014). Association between physical activity and mortality among breast cancer and colorectal cancer survivors: A systematic review and meta-analysis. *Annals of Oncology, 25*, 1293–1311.
30. Sprague, B. L., Trentham-Dietz, A., Egan, K. M., Titus-Ernstoff, L., Hampton, J. M., & Newcomb, P. A. (2008). Proportion of invasive breast cancer attributable to risk factors modifiable after menopause. *American Journal of Epidemiology, 168*, 404–411.
31. Xue, F., Willett, W. C., Rosner, B. A., Hankinson, S. E., & Michels, K. B. (2011). Cigarette smoking and the incidence of breast cancer. *Archives of Internal Medicine, 171*, 125–133.
32. Kushi, L. H., Doyle, C., McCullough, M., Rock, C. L., Demark-Wahnefried, W., Bandera, E. V., et al. (2012). American Cancer Society Guidelines on nutrition and physical activity for cancer prevention: Reducing the risk of cancer with healthy food choices and physical activity. *CA: A Cancer Journal for Clinicians, 62*, 30–67.
33. Kabat, G. C., Matthews, C. E., Kamensky, V., Hollenbeck, A. R., & Rohan, T. E. (2015). Adherence to cancer prevention guidelines and cancer incidence, cancer mortality, and total mortality: A prospective cohort study. *The American Journal of Clinical Nutrition, 101*, 558–569.
34. Cifu, G., & Arem, H. (2018). Adherence to lifestyle-related cancer prevention guidelines and breast cancer incidence and mortality. *Annals of Epidemiology, 28*, 767–773.
35. Harris, H. R., Bergkvist, L., & Wolk, A. (2016). Adherence to the World Cancer Research Fund/American Institute for Cancer Research recommendations and breast cancer risk. *International Journal of Cancer, 138*, 2657–2664.
36. Thomson, C.A., McCullough, M.L., Wertheim, B.C., Chlebowski, R.T., . Martinez, M.E., Stefanick, M.L. et al. Nutrition and physical activity cancer prevention guidelines, cancer risk, and mortality in the Women's Health Initiative. Cancer Prevention Research 2014, 7: 42–53.
37. Romaguera, D., Vergnaud, A.-C., Peeters, P. H., van Gils, C. H., Chan, D. S. M., Ferrari, P., et al. (2012). Is concordance with world cancer research fund/American Institute for Cancer Research guidelines for cancer prevention related to subsequent risk of cancer? Results from the EPIC study. *The American Journal of Clinical Nutrition, 96*, 150–163.
38. Hastert, T. A., Beresford, S. A., Patterson, R. E., Kristal, A. R., & White, E. (2013). Adherence to WCRF/AICR cancer prevention recommendations and risk of postmenopausal breast cancer. *Cancer Epidemiology, Biomarkers & Prevention, 22*, 1498–1508.
39. Kohler, L. N., Garcia, D. O., Harris, R. B., Oren, E., Roe, D. J., & Jacobs, E. T. (2016). Adherence to diet and physical activity cancer prevention guidelines and cancer outcomes: A systematic review. *Cancer Epidemiology, Biomarkers & Prevention, 25*, 1018–1028.
40. Schmid, D., & Leitzmann, M. F. (2014). Television viewing and time spent sedentary in relation to cancer risk: A meta-analysis. *Journal of the National Cancer Institute, 106*, 10.
41. Lipsky, L. M., & RIannotti, R. J. (2012). Associations of television viewing with eating behaviors in the 2009 health behaviour in school-aged children study. *Archives of Pediatrics & Adolescent Medicine, 166*, 465–472.
42. Harvie, M., Howell, A., & Evans, D. G. (2015). Can diet and lifestyle prevent breast cancer: what is the evidence? *American Society of Clinical Oncology Educational Book*, e66–e73.
43. Russo, J., & Russo, I. H. (1994). Toward a physiological approach to breast cancer prevention. *Journal of Cancer Epidemiology Biomarkers and Prevention, 3*, 353–364.
44. Russo, J. (2016). Reproductive history and breast cancer. *Hormone Molecular Biology and Clinical Investigation, 2016*(27), 3–10.

45. Cuzick, J. (2018). Progress in preventive therapy for cancer: a reminiscence and personal viewpoint. *British Journal of Cancer, 118*, 1155–1161.
46. Huang, Y., Fernandez, S., Goodwin, S., Russo, P. A., Russo, I. H., Sutter, T., & Russo, J. (2007). Epithelial to mesenchymal transition in human breast epithelial cells transformed by 17- beta- estradiol. *Cancer Research, 67*, 11147–11157.
47. Cuzick, J., & Baum, M. (1985). Tamoxifen and contralateral breast cancer. *Lancet, 2*, 282.
48. Fisher, B., Costantino, J. P., Wickerham, D. L., et al. (1998). Tamoxifen for prevention of breast cancer: Report of the National Surgical Adjuvant Breast and Bowel Project P-1 Study. *Journal of the National Cancer Institute, 90*, 1371–1388.
49. Powles, T. J., Ashley, S., Tidy, A., Smith, I. E., & Dowsett, M. (2007). Twenty-year follow- up of the Royal Marsden randomized, double-blinded tamoxifen breast cancer prevention trial. *Journal of the National Cancer Institute, 99*, 283–290.
50. Cuzick, J., Forbes, J., Edwards, R., et al. (2002). First results from the International Breast Cancer Intervention Study (IBIS-I): A randomized prevention trial. *Lancet, 360*, 817–824.
51. Cuzick, J., Forbes, J. F., Sestak, I., et al. (2007). Long-term results of tamoxifen prophylaxis for breast cancer–96-month follow-up of the randomized IBIS-I trial. *Journal of the National Cancer Institute, 99*, 272–282.
52. Veronesi, U., Maisonneuve, P., & Rotmensz, N. (2003). Italian randomized trial among women with hysterectomy: tamoxifen and hormone-dependent breast cancer in high-risk women. *Journal of the National Cancer Institute, 95*, 160–165.
53. Powles, T., Eeles, R., Ashley, S., et al. (1998). Interim analysis of the incidence of breast cancer in the Royal Marsden Hospital tamoxifen randomized chemoprevention trial. *Lancet, 352*, 98–101.
54. Cummings, S. R., Eckert, S., Krueger, K. A., et al. (1999). The effect of raloxifene on risk of breast cancer in postmenopausal women: Results from the MORE randomized trial. Multiple Outcomes of Raloxifene Evaluation. *Journal of the American Medical Association, 281*, 2189–2197.
55. Vogel, V. G., Costantino, J. P., Wickersham, D. L., et al. (2010). Update of the national surgical adjuvant breast and bowel project Study of Tamoxifen and Raloxifene (STAR) P-2 trial: Preventing breast cancer. *Cancer Prevention Research, 3*, 696–706.
56. Obiorah, I., & Jordan, V. C. (2011). Progress in endocrine approaches to the treatment and prevention of breast cancer. *Maturitas, 70*, 315–321.
57. Cuzick, J., Sestak, I., Cawthorn, S., Behalf of the IBIS-I Investigators, et al. (2015). Tamoxifen for prevention of breast cancer: Extended long-term follow-up of the IBIS-I breast cancer prevention trial. *The Lancet Oncology, 16*, 67–75.
58. Goss, P. E., Ingle, J. N., Ales-Martinez, J. E., et al. (2011). Exemestane for breast-cancer prevention in postmenopausal women. *The New England Journal of Medicine, 364*, 2381–2391.
59. Cuzick, J., Sestak, I., Forbes, J. F., et al. (2014). Anastrozole for prevention of breast cancer in high-risk postmenopausal women (IBIS-II): An international, double-blind, randomized placebo-controlled trial. *Lancet, 383*, 1041–1048.
60. Pagani, O., Regan, M. M., Walley, B. A., TEXT and SOFT Investigators; International Breast Cancer Study Group, et al. (2014). Adjuvant exemestane with ovarian suppression in premenopausal breast cancer. *The New England Journal of Medicine, 371*, 107–118.
61. Francis, P. A., Regan, M. M., Fleming, G. F., SOFT Investigators and the International Breast Cancer Study Group, et al. (2015). Adjuvant ovarian suppression in premenopausal breast cancer. *The New England Journal of Medicine, 372*, 436–446.
62. Litton, J. K., Arun, B. K., Brown, P. H., & Hortobagyi, G. N. (2012). Aromatase inhibitors and breast cancer prevention. *Expert Opinion on Pharmacotherapy, 13*, 325–331.
63. Subramani, R., & Lakshmanaswamy, R. (2017). Pregnancy and breast cancer. *Progress in Molecular Biology and Translational Science, 151*, 81–111.
64. Katz, T. A. (2016). Potential mechanisms underlying the protective effect of pregnancy against breast cancer: A focus on the IGF Pathway. *Frontiers in Oncology, 6*, 228.

65. Russo, J., Mailo, D., Hu, Y.-F., Balogh, G. A., Sheriff, F., & Russo, I. H. (2005). Breast differentiation and its implication in cancer prevention. *Clinical Cancer Research, 11*, 931s–936s.
66. Russo, J., Moral, R., Balogh, G. A., Mailo, D. A., & Russo, I. H. (2005). The protective role of pregnancy in breast cancer. *Breast Cancer Research Journal., 7*, 131–142.
67. Russo, J., Santucci-Pereira, J., López de Cicco, R., Sheriff, F., Russo, P. A., Peri, S., Slifker, M., Ross, E., MLS, M., Vidal, B. C., Belitskaya-Lévy, I., Arslan, A., Zeleniuch-Jacquotte, A., Bordas, P., Lenner, P., Ahman, J., Afanasyeva, Y., Hallmans, G., Toniolo, P., & Russo, I. H. (2012). Pregnancy-induced chromatin remodeling in the breast of postmenopausal women. *International Journal of Cancer, 131*, 1059–1070.
68. Santucci-Pereira, J., Zelenieuch-Jacquotte, A., Afanasyeva, Y., Zhong, H., Slifker, M., Peri, S., Ross, E. A., Lopez de Cicco, R., Zhai, Y., Nguyen, T., Sheriff, F., Russo, I. H., Su, Y., Arlsan, A. A., Bordas, P., Lenner, P., Ahman, J., Landstrom Eriksson, A. S., Johansson, R., Hallmans, G., Toniolo, P., & Russo, J. (2019). Genomic signature of parity in the breast of premenopausal women. *Breast Cancer Research, 21*, 46.
69. González-Jiménez, E., García, P. A., Aguilar, M. J., Padilla, C. A., & Álvarez, J. (2014). Breastfeeding and the prevention of breast cancer: a retrospective review of clinical histories. *Journal of Clinical Nursing, 23*, 2397–2403.
70. Tonetti, D. A. (2004). Prevention of breast cancer by recapitulation of pregnancy hormone levels. *Breast Cancer Research, 6*, E8.
71. Arun, B., Dunn, B. K., Ford, L. G., & Ryan, A. (2010). Breast cancer prevention trials: large and small trials. *Seminars in Oncology, 37*, 367–383.
72. Bodmer, M., Meier, C., Krahenbuhl, S., et al. (2010). Long-term metformin use is associated with decreased risk of breast cancer. *Diabetes Care, 33*, 1304–1308.
73. Chlebowski, R. T., McTiernan, A., Wactawski-Wende, J., et al. (2012). Diabetes, metformin, and breast cancer in postmenopausal women. *Journal of Clinical Oncology, 30*, 2844–2852.
74. van Staa, T. P., Patel, D., Gallagher, A. M., & de Bruin, M. L. (2012). Glucose-lowering agents and the patterns of risk for cancer: A study with the general practice research database and secondary care data. *Diabetologia, 55*, 654–665.
75. Decensi, A., Puntoni, M., Goodwin, P., et al. (2010). Metformin and cancer risk in diabetic patients: A systematic review and meta-analysis. *Cancer Prevention Research, 3*, 1451–1461.
76. Goodwin, P. J., Stambolic, V., Lemieux, J., et al. (2011). Evaluation of metformin in early breast cancer: A modification of the traditional paradigm for clinical testing of anticancer agents. *Breast Cancer Research and Treatment, 126*, 215–220.
77. Kim, J., Lim, W., Kim, E. K., et al. (2014). Phase II randomized trial of neoadjuvant metformin plus letrozole versus placebo plus letrozole for estrogen receptor positive postmenopausal breast cancer (METEOR). *BMC Cancer, 14*, 170.
78. Khan, Q. J., Kimler, B. F., Reddy, P. S., et al. (2012). Randomized trial of vitamin D3 to prevent worsening of musculoskeletal symptoms in women with breast cancer starting adjuvant letrozole: the VITAL trial. *Journal of Clinical Oncology, 30*, abstr 9000.
79. Santa-Maria, C. A., & Stearns, V. (2013). Statins and breast cancer: Future directions in chemoprevention. *Current Breast Cancer Report, 5*, 161–169.
80. Yang, B., Ren, X. L., Fu, Y. Q., Gao, J. L., & Duo, L. (2014). Ratio of n-3/n-6 PUFAs and risk of breast cancer: A meta-analysis of 274,135 adult females from 11 independent prospective studies. *BMC Cancer, 14*, 105.
81. Fabian, C. J., & Kimer, B. F. (2013). Marine-derived omega-3 fatty acids: fishing for clues for cancer prevention. *American Society of Clinical Oncology Educational Book*, 97–101.
82. Lyengar, N. M., Hudis, C. A., & Gucalp, A. (2013). Omega-3 fatty acids for the prevention of breast cancer: An update and state of the science. *Current Breast Cancer Reports, 5*, 247–254.
83. National Comprehensive Cancer Network. Clinical practice guidelines in oncology: Breast cancer: Risk reduction, Version 2. 2014. Available at: http://www.nccn.org/professionals/physician_gls/f_guidelines.asp#detection. Accessed 1 Jan 2015.

84. Ford, D., Easton, D. F., Bishop, D. T., et al. (1994). Risks of cancer in BRCA1 mutation carriers. *Lancet, 343*, 692–695.
85. Metcalfe, K., Lynch, H. T., Ghadirian, P., et al. (2004). Contralateral breast cancer in BRCA1 and BRCA2 mutation carriers. *Journal of Clinical Oncology, 22*, 2328–2335.
86. Metcalfe, K. A., Lubinski, J., Ghadirian, P., et al. (2008). Predictors of contralateral prophylactic mastectomy in women with a BRCA1 or BRCA2 mutation: The hereditary breast Cancer clinical study group. *Journal of Clinical Oncology, 2*, 1093–1097.
87. McDonnell, S. K., Shaid, D. J., Myers, J. L., et al. (2001). Efficacy of contralateral prophylactic mastectomy in women with a personal and family history of breast cancer. *Journal of Clinical Oncology, 19*, 3938–3943.
88. Meijers-Heijboer, H., van Geel, B., van Putten, W. L., et al. (2001). Breast cancer after prophylactic bilateral mastectomy in women with a BRCA1 or BRCA2 mutation. *The New England Journal of Medicine, 345*, 159–164.
89. Hartmann, L. C., Sellers, T. A., Schaid, D. J., et al. (2001). Efficacy of bilateral prophylactic mastectomy in BRCA1 and BRCA2 gene mutation carriers. *Journal of the National Cancer Institute, 93*, 1633–1637.
90. Rebbeck, T. R., Friebel, T., Lynch, H. T., et al. (2004). Bilateral prophylactic mastectomy reduces breast cancer risk in BRCA1 and BRCA2 mutation carriers: The PROSE Study Group. *Journal of Clinical Oncology, 22*, 1055–1062.
91. Van Sprundel, T. C., Schmidt, M. K., Rookus, M. A., et al. (2005). Risk reduction of contralateral breast cancer and survival after contralateral prophylactic mastectomy in BRCA1 or BRCA2 mutation carriers. *British Journal of Cancer, 8*, 287–292.
92. Heemskerk-Gerritsen, B., Brekelmans, C., Menke-Pluymers, M., et al. (2007). Prophylactic mastectomy in BRCA1/2 mutation carriers and women at risk of hereditary breast cancer: Long-term experiences at the Rotterdam Family Cancer Clinic. *Annals of Surgical Oncology, 14*, 3335–3344.
93. Metcalfe, K., Gershman, S., Ghadirian, P., et al. (2014). Contralateral mastectomy and survival after breast cancer in carriers of BRCA1 and BRCA2 mutations: Retrospective analysis. *BMJ, 348*, 226.
94. Burke, E. E., Portschy, P. R., & Tuttle, T. M. (2014). Contralateral prophylactic mastectomy: Are we overtreating patients? *Expert Review of Anticancer Therapy, 14*, 491–494.
95. Portschy, P. R., Kuntz, K. M., & Tuttle, T. M. (2014). Survival outcomes after contralateral prophylactic mastectomy: A decision analysis. *Journal of the National Cancer Institute, 106*, 160.
96. Eisen, A., Lubinski, J., Klijn, J., et al. (2005). Breast cancer risk following bilateral oophorectomy in BRCA1 and BRCA2 mutation carriers: An international case-control study. *Journal of Clinical Oncology, 20*, 7491–7496.
97. Domchek, S. M., Friebel, T. M., Singer, C. F., et al. (2010). Association of risk-reducing surgery in BRCA1 or BRCA2 mutation carriers with cancer risk and mortality. *Journal of the American Medical Association, 304*, 967–975.
98. Rebbeck, T. R., Levin, A. M., Eisen, A., et al. (1999). Breast cancer risk after bilateral prophylactic oophorectomy in BRCA1 mutation carriers. *Journal of the National Cancer Institute, 91*, 1475–1479.
99. Metcalfe, K., Gershman, S., Lynch, H. T., et al. (2011). Predictors of contralateral breast cancer in BRCA1 and BRCA2 mutation carriers. *British Journal of Cancer, 26*, 1384–1392.
100. Metcalfe, K., Lynch, H. T., Ghadirian, P., et al. (2011). Risk of ipsilateral breast cancer in BRCA1 and BRCA2 mutation carriers. *Breast Cancer Research and Treatment, 127*, 287–296.
101. World Cancer Research Fund International. Link between lifestyle and cancer. Available at: wcrf.org/int/cancer-facts-figures/link-between-lifestyle-cancer-risk.

Chapter 8
The Physiological Basis of Breast Cancer Prevention

Abstract Epidemiological data from various parts of the world have consistently shown that early full-term pregnancy and multiparity are associated with breast cancer risk reduction in postmenopausal women, whereas late pregnancy and nulliparity are associated with increased risk. Our laboratory has provided evidence supporting the concept that the degree of differentiation acquired through an early pregnancy changes the genomic signature of the lobular structures of the parous breast from those of the nulliparous one. Altogether, parous premenopausal and postmenopausal breast show a specific transcriptome profile, in which genes controlling chromatin remodeling and cell differentiation are activated after FTP and stay upregulated for many years and are the hallmark of the postmenopausal breast.

Keywords Early full-term pregnancy · Multiparity · Breast cancer risk · Postmenopausal women · Nulliparity · Transcriptome profile · Chromatin remodeling · Cell differentiation · p53 · BRCA1

8.1 Breast Cancer Prevention

Breast cancer is a heterogeneous and complex disease resulting from the uncontrolled growth of cells that are unique and specific to the breast. The occurrence of cancer of the breast has long been known, as documented in the Edwin Smith surgical papyrus, written between 3000 and 1500 BC (Table 8.1). A common denominator for the risk of developing breast cancer has been found to be a reproductive history. Increased breast cancer incidence and mortality were associated with nulliparity as early as the 1700s, as reported by Bernardino Ramazzini, who attributed the phenomenon to the childlessness of nuns in Italian convents (Table 8.1).

Epidemiological data from various parts of the world have consistently shown that early full-term pregnancy and multiparity are associated with breast cancer risk reduction in postmenopausal women [6–8], whereas late pregnancy and nulliparity are associated with increased risk [3]. It has been postulated that the mechanism of pregnancy-induced protection is mediated by changes in environmental settings [9], and/or alterations in the immunological profile of the host [10]. Our studies on the differentiation of the breast [4, 5, 11–13] under the influence of the complex

J. Russo, *The Future of Prevention and Treatment of Breast Cancer*,
https://doi.org/10.1007/978-3-030-72815-1_8

Table 8.1 Landmarks in the history of breast cancer

Year	Landmark discovery or observation	Reference
3000–1500 BC	Edwin Smith surgical papyrus describes eight cases of breast cancer in women	Breasted [1]
1700	Bernardino Ramazzini attributed the high incidence of breast cancer in nuns to their childlessness	Mustacchi [2]
1970	MacMahon et al. reported that pregnancy exerted a protective effect in women whose first child was born from early teens to the middle 20s	MacMahon et al. [3]
1978	Russo and Russo established the concept the mammary gland differentiation induced by full-term pregnancy is the mechanism behind the protective effect against breast cancer	Russo and Russo [4, 5]

hormonal milieu created by two newly formed endocrine organs, the placenta and the fetus [14], have unraveled the morphological, functional, genomic, and transcriptomic changes that ultimately result in the induction of a permanent and specific profile that serves as an indicator of reduced cancer risk [15, 16]. My laboratory has provided evidence supporting the concept that the degree of differentiation acquired through an early pregnancy changes the genomic signature that differentiates the lobular structures of the parous breast from those of the nulliparous one [4, 5, 15]. With pregnancy and lactation, the mammary parenchyma reaches the final stage of secretory lobule type 4 (Lob 4) that forms by the end of the reproductive process and remains present during lactation until they involute to Lob 1 in postmenopausal years. Therefore, the protection conferred by an early pregnancy is mediated by the induction of differentiation that we defined as the coordinated and sequential series of events induced in the breast by the hormonal milieu of pregnancy, or pregnancy-like conditions (such as hCG administration), which culminate in the activation of genes controlling ductal and lobular development inducing a unique genomic signature. This genomic signature is induced by the process of lobular differentiation that takes place when the Lob1 differentiates in Lob 2 and then in Lobs 3 and 4. Therefore the Lob 1 of the involute breast of parous women, although it looks morphologically similar to the Lob 1 of nulliparous women, is different because the latter has undergone the process of full differentiation where the genomic changes have been imprinted [4, 5, 15, 16]. One way to assess whether a specific genomic fingerprint is permanently imprinted in the breast by a full-term pregnancy (FTP) is to compare the transcriptomic profiles of breasts from parous and nulliparous women. For this purpose, we decided to use a genome-wide approach to identify long-term genomic changes associated with FTP by studying breast core needle biopsies (CNBs) obtained from healthy pre- and postmenopausal volunteers [17].

8.2 Genomic Signature of the Breast Changes After First Full-Term Pregnancy in Premenopausal Women

In a study published by our laboratory [18], we studied the parity-associated gene expression differences in the breast of premenopausal women. Because the breast undergoes involution after full-term pregnancy (FTP), we examined the gene expression differences in parous (P) vs. nulliparous women (NP) as a function of time since last FTP. The study subjects were volunteers recruited among healthy women between the ages of 29 and 47 years and residing in Norrbotten County, Sweden [18]. Samples from 109 women (30 NP and 79 P) were included in the main study comparing P vs. NP. Using the discovery and validation approach, we found [18] 54 probesets, representing 43 genes, that were differentially expressed between P and NP women. Of the 43 genes, 40 were upregulated and 3 downregulated in the parous premenopausal breast [18]. Upregulated genes in the P breast included APOBEC3G, DSC3, FZD8, and EAF2, while FOXQ1 was among the downregulated genes. Analyses of gene expression according to the time since last pregnancy (TSLP) identified 286 genes (416 probesets) differentially expressed between P and NP samples (Fig. 8.1).

For probesets upregulated in the P women, gene expression differences are clustered into three expression patterns as a function of TSLP (Fig. 8.2). The first cluster consisted of 83 genes (107 probesets) which were upregulated following the last FTP, but whose expression progressively returned to the level of expression observed in NP women (Fig. 8.2a). These 83 genes were named "transient" genes. The second cluster consisted of probesets for which the P-NP differences were the highest for women with <5 years since last FTP, decreased with increasing TSLP but remained elevated as compared to NP women. The 95 genes (154 probesets) in this cluster were called "long term changing" (Fig. 8.2b). In the last cluster, which included 60 genes (91 probesets), the fold changes between P an NP appeared constant, regardless of the TSLP. Therefore, we called these "long-term constant" genes (Fig. 8.2c).

Gene Ontology (GO) enrichment analyses showed an abundance of GO terms associated to developmental processes or immune response (Fig. 8.3); other less-abundant enriched GO terms were related to proliferation, intracellular transport, and cell death. Among the genes whose expression was transiently affected by parity 55%, including CD8A, XCL1, and GZMA [18], enriched GO terms related to immune response. Among the genes classified as long-term changing, 32% were related to immune response (e.g., CD2), and 24% were involved in developmental processes (e.g., EAF2). Notably, of the long-term constant genes, 56% were related to developmental/differentiation processes, including EGR3, DSC3, KRT5, and FZD8. These data indicate that the proportion of genes involved in immune response decreases among the genes that are upregulated irrespectively of TSLP, while the proportion of genes related to developmental processes increases (chi-squared p value = $p < 0.001$).

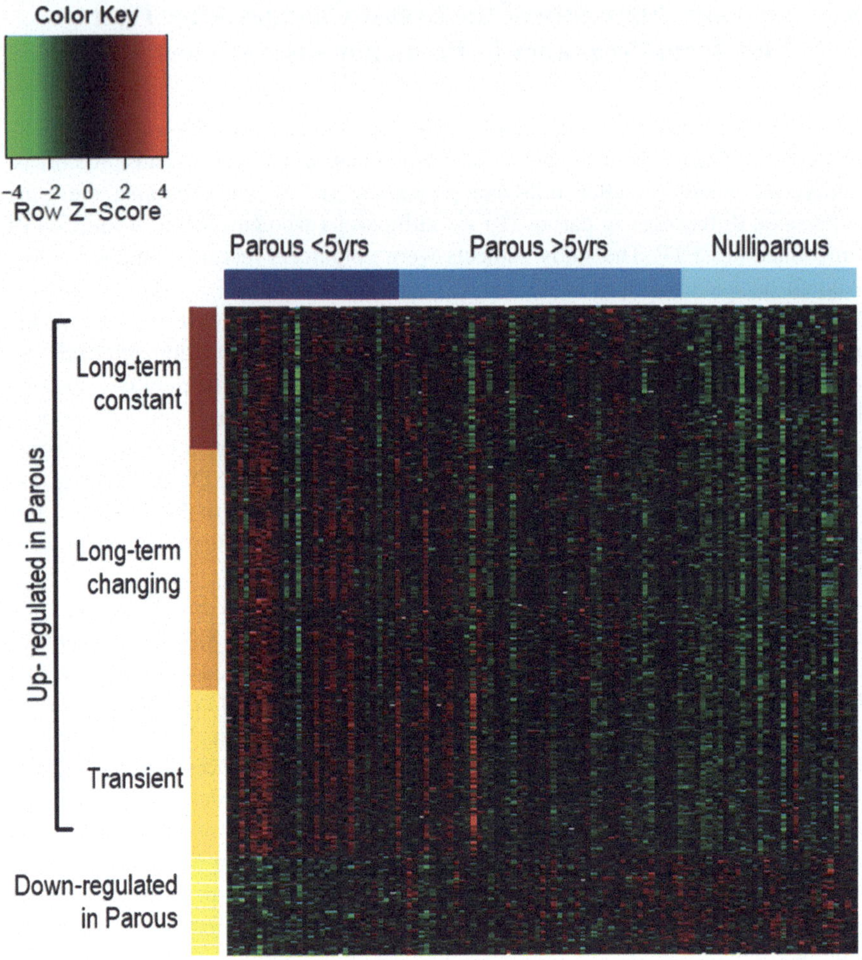

Fig. 8.1 Heatmap of all 416 probesets found differentially expressed between parous and nulliparous. Each column represents a volunteer. Dark blue represents parous women within 5 years since last full-term pregnancy, blue represents parous women with more than 5 years since pregnancy, and cyan represents nulliparous women. Each line corresponds to a probeset, which, accordingly with its pattern of expression within the parous breast samples compared to nulliparous, was classified as downregulated (light yellow), transiently upregulated (yellow), long-term changing (orange), and long-term constant upregulated (brown). (Reprinted from: Santucci-Pereira et al. [18])

Evaluation of the canonical pathways represented by the differentially expressed genes also showed that pathways involved in signaling and activation of T cells were enriched by both transient and long-term genes (Fig. 8.4).

The 48 genes (64 probesets) downregulated in the P breast fell into two patterns. Twenty-two genes, including WIF1, EDN1, CXCL1, and FOXQ1, were downregulated in the P breast and remained with lower expression levels compared to NP breast irrespective of the number of years since last FTP. The remaining 26 genes

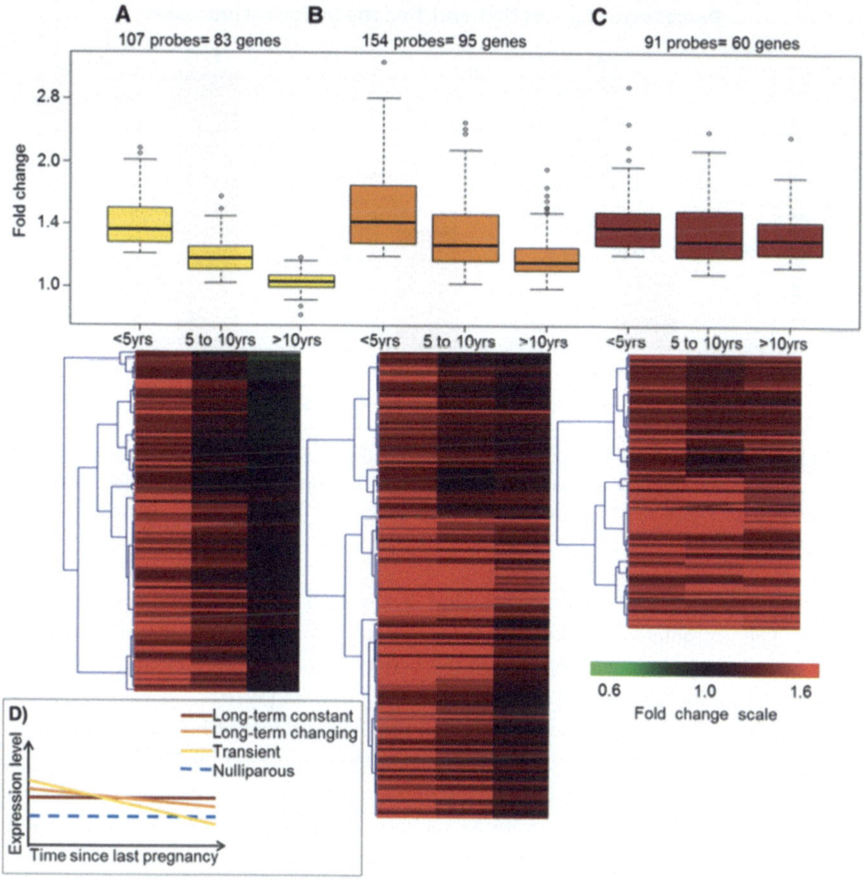

Fig. 8.2 Three expression patterns were identified among the upregulated genes. In the top, the graph shows the fold change ranges of each cluster: (**a**) transient, (**b**) long-term changing, and (**c**) long-term constant according to time since last full-term pregnancy (<5 years, 5–10 years, and > 10 years). Heatmaps show the expression patterns of each cluster. (**d**) Schematic representation of the expression levels of each cluster in function of time since last pregnancy. (Reprinted from: Santucci-Pereira et al. [18])

were downregulated in women with TSLP <5 years; however, the expression increased reaching similar or higher levels than those in NP women when TSLP increased. These genes included DLG5, KDM4B, and TOX3. We did not conduct a GO enrichment analysis for these genes because of their limited number [18].

Finally, we examined genes reported to be related to chromatin remodeling. Among those upregulated, we observed APOBEC3G, TOX, UHRF1, and NAP1L2, while KDM4B and TOX3 were downregulated.

P/NP differences in gene expression were confirmed for 8 out of the 11 genes analyzed by real-time PCR in a subset of 37 samples (20 P and 17 NP): Ct values, ΔCts, and −ΔΔCts. WIF1 was downregulated, while EAF2, BHLHE22,

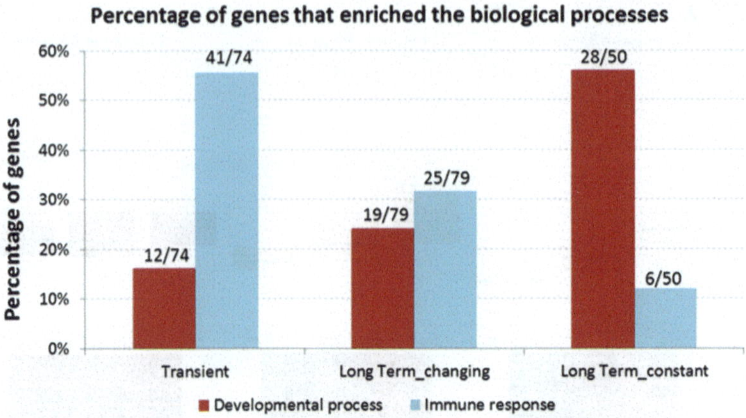

Fig. 8.3 Graph shows the percentage of genes associated with GO terms related to developmental processes (red) and immune response (blue) for each of the three clusters. 74, 79, and 50 are the total number of genes associated at least with one GO term for each cluster. (Reprinted from: Santucci-Pereira et al. [18])

APOBEC3G, DSC3, KRT5, EGR3, and RASGRP1 were upregulated in the P breast. The intraclass correlation coefficient (ICC) comparing the real-time to microarray measurements varied from 0.35 (EAF2) to 0.92 (WIF1), and Spearman correlation coefficients varied from 0.41 (EAF2) to 0.94 (WIF1).

We also analyzed the 11 selected genes in 5796 patients of breast cancer available in cBioPortal. One or more of these genes were altered in 770 (13%) breast cancer patients. BHLHE22 was the gene that appears altered in the largest percentage of patients (6.9%); the alterations included amplification, deep deletion, and missense mutations. The other ten genes showed to be altered in 1.5% or less patients. An overall Kaplan–Meier survival curve has been generated comparing the group of patients with these 11 genes altered versus patients without the alterations. The group of patients that contain alteration in these genes has shorter overall survival, median of 143.13 versus 168.3 months, logrank test p value = 9.09e−5 (Fig. 8.5).

We performed IHC analyses to evaluate whether the changes in gene expression reflected into the protein expression levels. We have evaluated two proteins related to differentiation Desmocollin 3 and Cytokeratin 5. These two markers were presented in all tested samples, and although we have observed slight upregulation of their genes by microarray and RT-PCR, we did not detect protein differences between P and NP breast tissues. Related to immune response, we evaluated markers for dendritic cells because of their role in antigen presentation for T cell activation. Using two markers CD123 and LAMP3, we detected few positive cells with no differences between P and NP samples. We also evaluated CD2 and CD3D, markers of T cell activation (Fig. 8.6.). The percentage of positive cells for CD2 was low in both groups; we observed slight increase in the percentage of CD2-positive cells in parous women (1.6 times; P median = 0.59%; NP median = 0.31%); however, it was

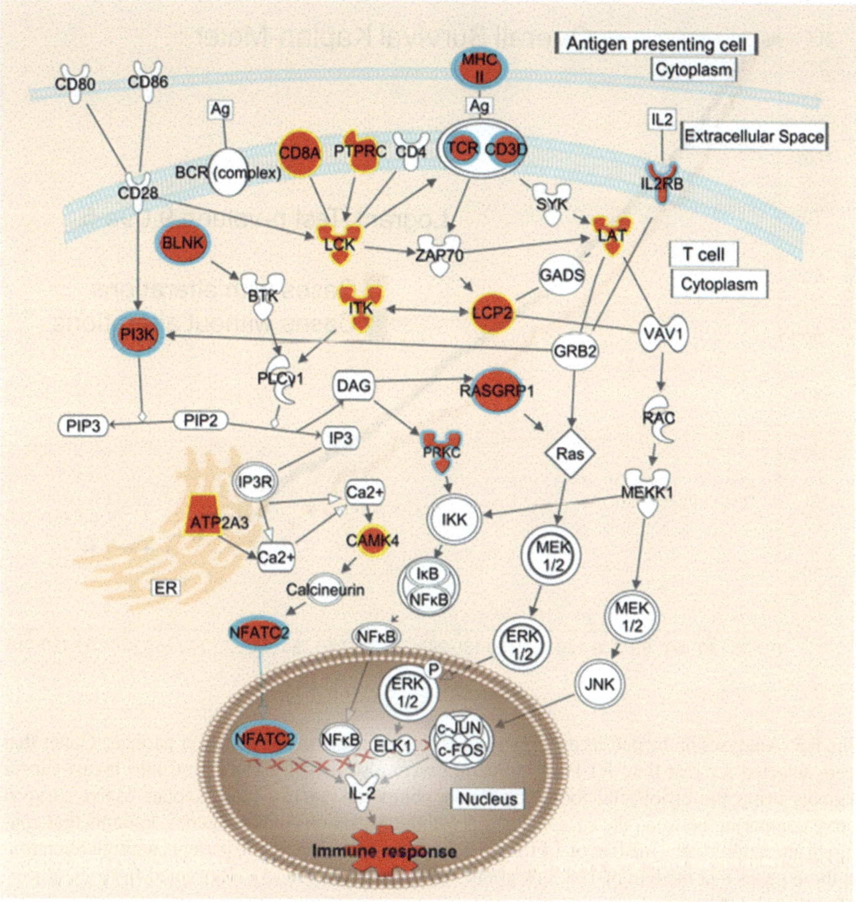

Fig. 8.4 Pathway representing T cell signaling. Genes in red were upregulated in the parous breast. Genes with a blue border had a long-term effect, while genes with a yellow border were transiently upregulated. Pathway was modified from IPA software (QIAGEN) using Path Designer. (Reprinted from: Santucci-Pereira et al. [18])

not statistically significant (p value = 0.156). The overall percentage of CD3D-positive cells was higher ranging from 0.6% to 9.4%. We confirmed a statistically significant twofold increase in the percentage of CD3D-positive cells in the P breast (P median = 3.28%; NP median = 1.62%; p value = 0.006) (Fig. 8.6). We also compared whether the percentage of positive cells between P women with TSLP ≤ 5 and > 5 differed. For both CD2 and CD3D, we see a slightly larger number of positive cells in TSLP ≤ 5, but were not statistically significant (CD2 p value = 0.16 and CD3D p value = 0.09).

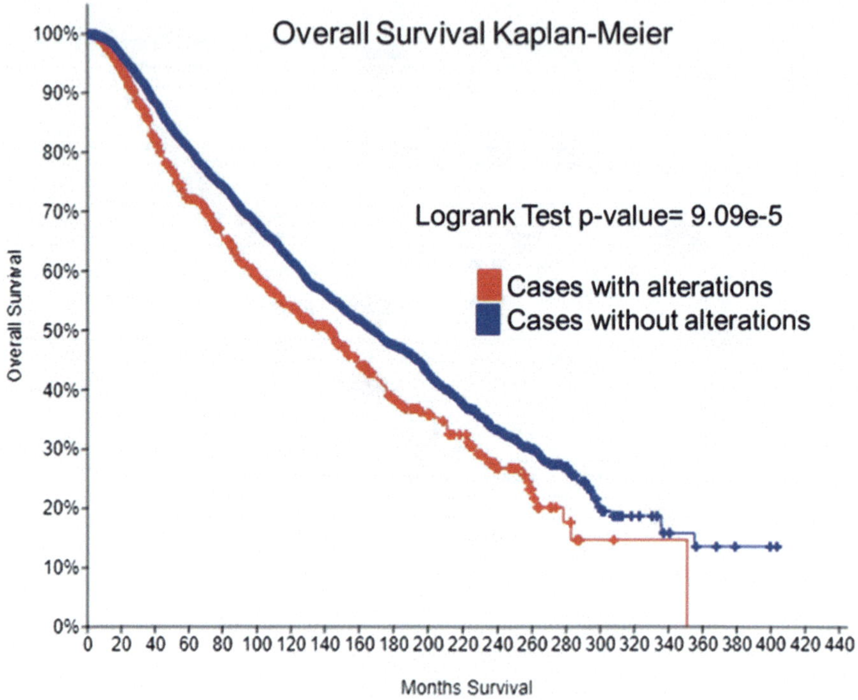

Fig. 8.5 Analyses of the parous differentially expressed genes in breast cancer patients. Genes that were selected for real-time RT-PCR validation were also used for evaluation into breast cancer patients using the cBioPortal tool (http://www.cbioportal.org/), overall Kaplan–Meier survival curve comparing between the cases with and without alterations in these genes. Patients that contained alterations had a median of 143.13 months overall survival, while patients without alteration in these genes had median of 168.3 (logrank test p value = 9.09e−5). (Reprinted from: Santucci-Pereira et al. [18])

8.3 Biological Significance of the Genomic Changes Post Full-Term Pregnancy in Premenopausal Women

We have evaluated gene expression differences in parous and non-parous breast using biopsies from healthy premenopausal volunteers. Using a discovery/validation approach, we found 43 differentially expressed genes [18]. Evaluation of expression differences between NP and P as a function of TSLP (Time since last pregnancy) identified 238 genes up- and 48 genes downregulated in the P breast. The downregulated genes fell into two patterns of expression (transient and long-term), while the upregulated genes fell into three patterns (transient, long-term changing, and long-term constant) (Fig. 8.2). Through GO enrichment analyses, we found that genes whose expression was transiently increased after pregnancy were mainly related to immune response. Long-term changing genes included immune- and development-related genes, while genes categorized as long-term constant were mainly involved in cell differentiation and developmental processes (Fig. 8.3).

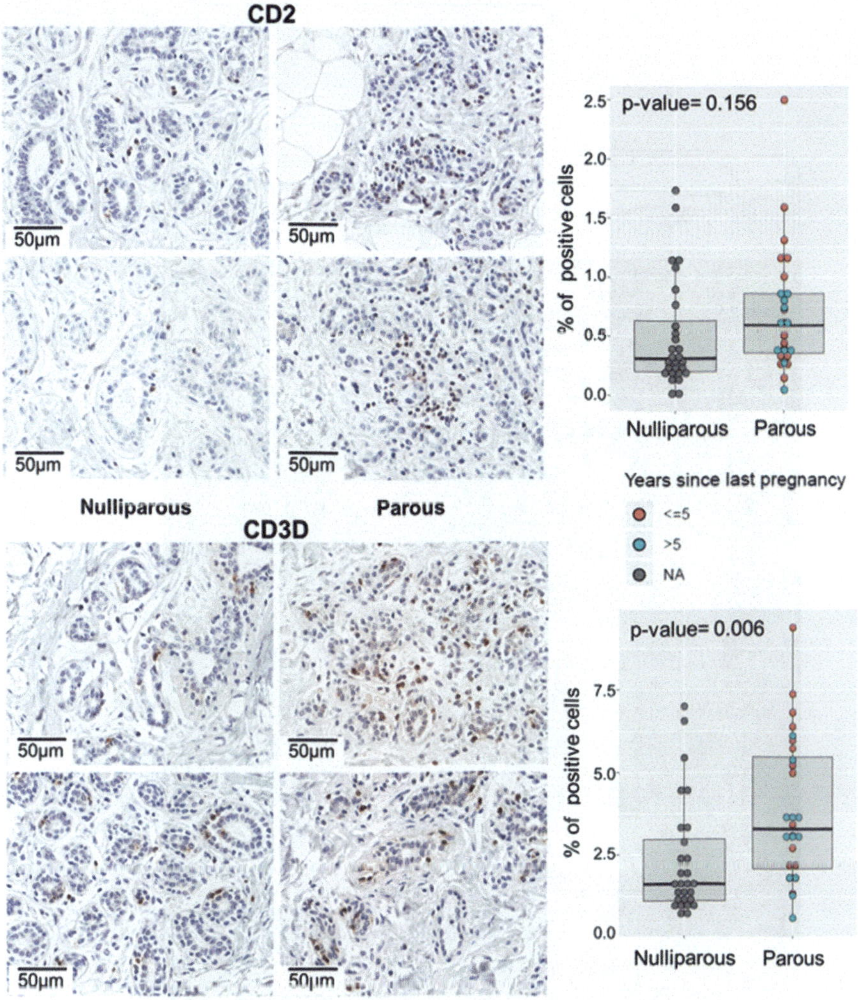

Fig. 8.6 Immunohistochemical evaluation of CD2 and CD3D for T cell activation. On the top are results regarding CD2, and on the bottom are results of CD3D. Panels on the left show the immunohistochemical reactions (×400); each panel contains two nulliparous samples (left) and two parous samples (right). On the right, boxplots show the percentage of positive cells. We observe an increase in the percentage of CD3D-positive cells in the parous group (*p* value = 0.006). (Reprinted from: Santucci-Pereira et al. [18])

8.4 The Genomic Signature of the Postmenopausal Breast

Samples from 113 post-menopausal women ranging from 49 to 69 years of age were included in this study comparing 42 NP vs 71 P women, using the discovery and validation approach described in [17, 19, 20]. We have identified 305

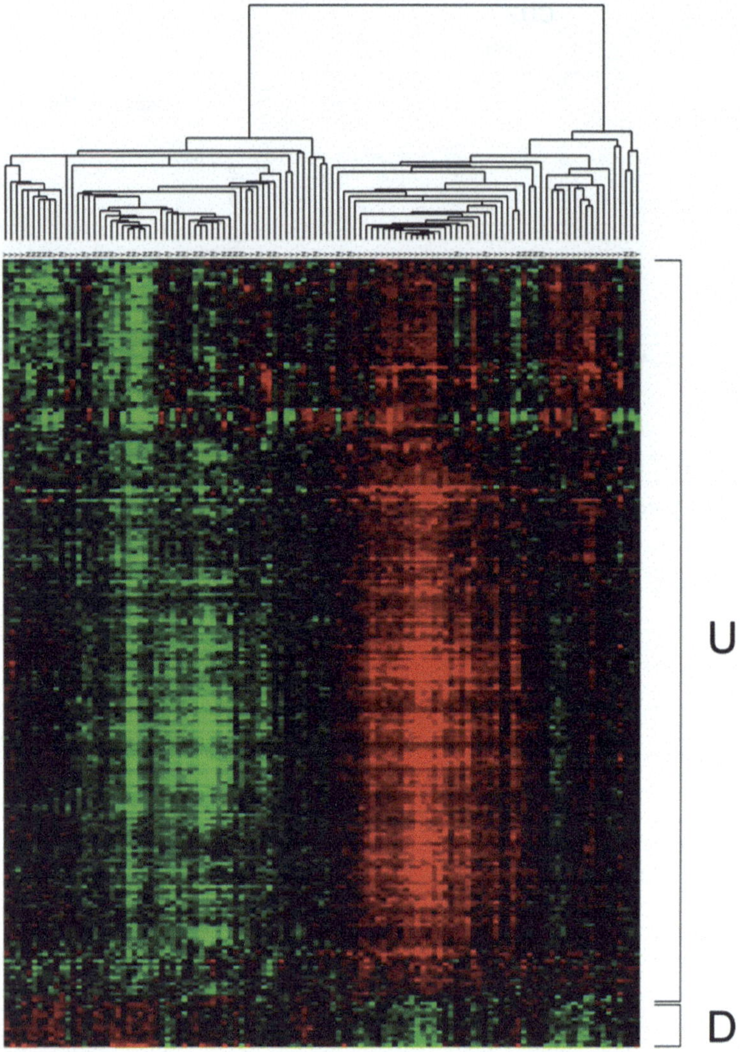

Fig. 8.7 Hierarchical clustering of differentially expressed probesets in parous and nulliparous women. Red represents expression values above the median across all samples, and green represents values below the median. In the two top-level clusters of samples, the right cluster is composed mainly of parous samples, and the left cluster is composed mainly of nulliparous samples. *U* represents the intensity of upregulated probesets among parous samples, whereas *D* represents the intensity of downregulated probesets. *N* represents nulliparous and *Y* represents parous. A chi-square test of independence on parity status and sub-tree membership resulted in a *p*-value of 0.001. (Reprinted from: Peri et al. [20])

differentially expressed probesets (corresponding to 208 distinct genes) between P and NP women. Of these, 267 were upregulated and 38 were downregulated (Fig. 8.7). To understand the biological theme of the observed gene expression differences, we carried out bioinformatics-based analysis of microarray data. Gene ontology (GO) enrichment analysis revealed biological processes that were categorized into groups including RNA metabolic processes, differentiation and development of epidermis and ectoderm, and cell-substrate junction assembly, findings that are in agreement with existing knowledge that pregnancy hormones promote the differentiation of mammary epithelial cells [17–20]. Highly represented in the parous breast were biological processes involving both mRNA and RNA metabolic processes and RNA splicing machinery. Biological processes such as differentiation and development of epithelial and ectodermal cells as well as genes which are pivotal in anchoring of epithelial cells to the basement membrane, hemidesmosome, and cell-substrate junction assembly were upregulated in the P group. More importantly was the finding that genes controlling the process of chromatin remodeling were also involved in the changes induced by pregnancy and that are pivotal for the long protective effect of early pregnancy against breast cancer [19]. Among the downregulated genes, insulin-like growth factor 1 (IGF-1) was enriched in 19 biological processes that comprised cell proliferation, regulation of IGF-like growth factor receptor signaling, somatic stem cell maintenance, muscle cell differentiation, and apoptosis, among others [19, 20].

One of the most exciting parts of this study was identifying the genes that were up- and downregulated by an early pregnancy and, most important, how they cluster together and deciphering the possible mechanism of its action. It is in this way that we discovered the complex molecular machine found in the speckles of the nuclei were associated with the splicing mechanism of the eukaryotic cells [21] (Fig. 8.8).

The pre-mRNA splicing factors are enriched in nuclear domains termed interchromatin granule clusters or nuclear speckles. Among the members of the splicing factor compartment are CCNL1 and CCNL2, which participate in the pre-mRNA splicing process and are located in the nuclear speckles [23, 24]. CCNL1 and CCNL2 are upregulated in the parous breast, and the protein of the latter is also overexpressed in the nucleus of breast epithelial cells [19]. CCNL1 and CCNL2 are

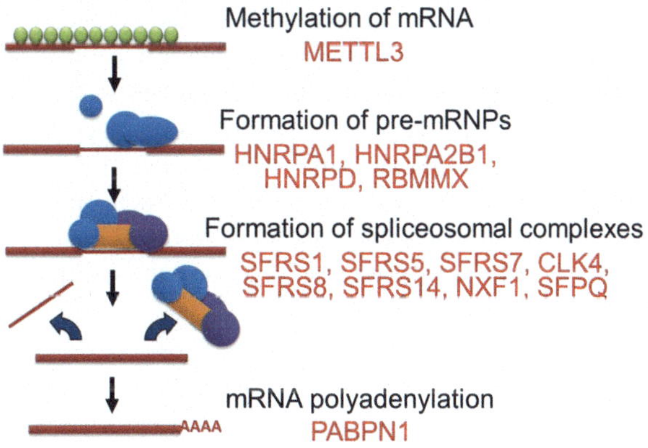

Fig. 8.8 Scheme of RNA processing pathway and up-regulated genes in parous breast tissue [22]

transcriptional regulators [23, 24] that modulate the expression of critical factors leading to cell apoptosis, possibly through the Wnt signal transduction pathway [25], a signaling pathway that is enriched by downregulated genes in the parous breast.

In published preclinical and clinical studies [8, 11, 13, 15], it has been reported that pregnancy confers protection from breast cancer development by inducing gland differentiation, which imprints a specific and permanent genomic signature in this organ. To reemphasize this concept, we compared the study reported by us in 2008 [8], a case-control study of transcriptomic analysis of normal breast tissues obtained from parous and nulliparous women free of breast pathology and parous and nulliparous women with a history of breast cancer who served as controls and cases, respectively, with the study that we performed several years later with another population of women and reported in 2012 [20]. The relevance of this comparison is that when we applied GO enrichment analysis to the gene lists generated from both studies, it was found that in both studies the processes involved in RNA metabolism and RNA processing were similar. This uniqueness of the data push us forward to study this mechanism in more detail, and for that reason we performed RNA-sequencing P and NP breast biopsies from postmenopausal women free of disease. Alternative splicing events were identified through the Multivariate Alternative Transcript Splicing Software (MATS) and considered statistically significant when the FDR (false discovery rate) was less than 5 percent. We evaluated the differences between P and NP of five types of splicing: mutually exclusive exon, alternative 5′ splice site, alternative 3′ splice site, skipped exon, and retained intron. The last type, the retained intron, had the higher number of statistically significant events different between P and NP. From 95 retained intron (RI) events, P women had 21 events, while in the NP we found 74 RI events. Retained intron (RI) events were selected for further examination because these events can either be splicing errors or a form of gene expression regulation, which indicated that they have lower expression and/or they produce lower quantities of functional proteins. Several of the genes containing RI belong to PI3K-Akt signaling and spliceosome pathways. Gene ontology (GO) analysis revealed that those genes with higher levels of RI in the NP breast enriched mainly GO terms related to RNA processing and splicing. This is in agreement with our previous gene expression results, in which genes involved in RNA processing and splicing were downregulated in the NP women. The relevance of this data is that alternative splicing is disproportionately affecting important pathways such as PI3K-Akt, spliceosome, and other genes involved in mRNA processing in NP women which may explain their higher risk for breast cancer. These data led us to suggest that post-transcriptional regulation is a key element in the differences between the breasts of parous and nulliparous women and their lifetime breast cancer risk [22].

8.4.1 The Role of Long Noncoding RNA in the GSP of the Menopausal Breast

In our study we were able to see few sequences identified as long noncoding RNA. This data opened our minds to a new set of mechanisms that had never been explored as relevant since noncoding RNAs were once thought of as the "dark matter" of the genome, but it is becoming increasingly clear that they play major roles in gene regulation as 98 percent of the human genome is non-protein-coding material. Noncoding RNAs are transcripts of RNA that do not code for a protein and can be classified as short noncoding RNA (e.g., miRNA; between 18 and 22 nucleotides in length) and long noncoding RNA (lncRNA; equal or greater than 200 nucleotides in length). Long noncoding RNAs have diverse gene expression regulatory functions including transcriptional regulation, post-transcriptional regulation, or direct regulation of proteins, and they may act as tumor suppressors while others act as oncogenes.

The noncoding RNAs that include XIST, MALAT-1 (also called NEAT2), and NEAT1 are upregulated in the parous breast. XIST, which inactivates X chromosome as an early developmental process, plays an essential role in female mammals by providing dosage equivalence between males and females. Upregulation of XIST occurs upon differentiation, whereas failure to express XIST is often seen in malignancies and in early embryogenesis [26]. Our findings are supported by reports in the literature suggesting that XIST is expressed in adult well-differentiated cells in order to maintain gene repression [26–29].

Oxytocin, a neurotransmitter that acts through its specific receptor OXTR and is overexpressed during lactation, upregulates the expression of MALAT-1, a highly conserved noncoding RNA [30–33]. Interestingly, both MALAT1 and OXTR remain overexpressed in the breast of postmenopausal parous women.

NEAT1 and NEAT2 localize to the periphery and to the interior the spliceosome assembly factor SC35 domains, or speckles. Our observation [19] that in breast epithelial cells CCNL2 is highly enriched in nuclear speckles indicates that CCNL2 might co-localize with NEAT1 and NEAT2. The downregulation of NEAT1, NEAT2, and XIST in the breast of nulliparous women, in whom this organ never reached a stage of complete differentiation similar to that achieved after completion of pregnancy and lactation [19, 20], suggests that the undifferentiated breast is not actively involved in the RNA metabolism that is necessary for maintaining a state of differentiation. In a study performed in our laboratory by Dr. Maria Barton [34], it was found 42 new lncRNAs that were differentially expressed between the NP and parous menopausal women (Fig. 8.9).

As exciting was to have so much unexpected differences, we started first to search the literature to identify lncRNAs involved in mammary gland development and/or disease, and no literature was found detailing any of our specific lncRNAs (except for lncEPCAM which was vaguely described as BC200), so we determined that these were all novel lncRNAs in our system. We further tried to determine if some of these regions may have aliases depending on the lncRNA database used,

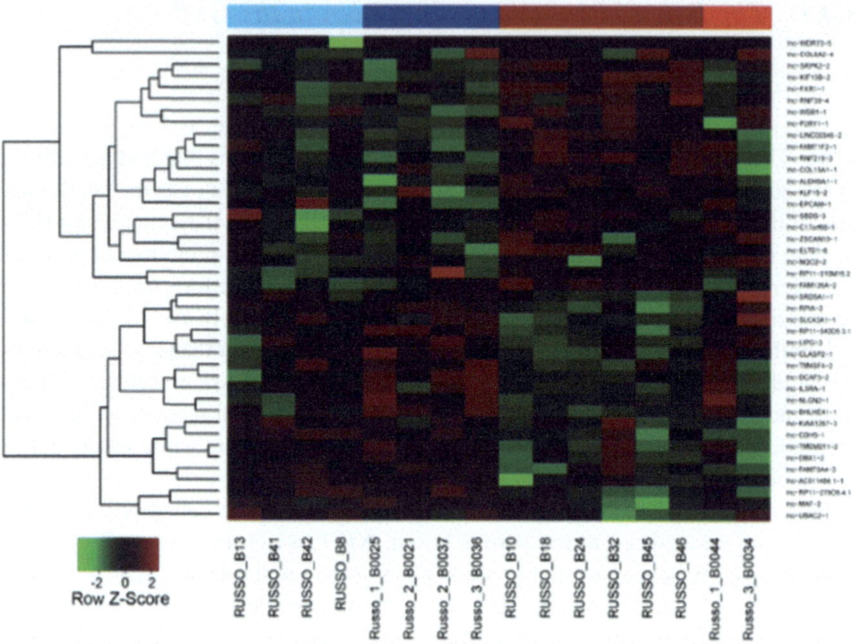

Fig. 8.9 Heatmap representing differentially expressed lncRNAs between parous and nulliparous women. Upregulated lncRNAs are represented in red, and downregulated lncRNAs are represented in green. Light and dark blue colors under parous represent two different batches of samples. Both batches were determined to be comparable, and thus samples were grouped under "parous." Idem for dark and bright red under nulliparous [34]

and for this purpose we used LNCipedia [https://www.ncbi.nlm.nih.gov/pubmed/23042674], which is a comprehensive compendium of human lncRNAs. We next analyzed the coverage of earlier sequencing with the software tool called Integrated Genomics Viewer (IGV) [https://www.ncbi.nlm.nih.gov/pmc/articles/PMC3346182/] in order to ensure that custom probes qPCR would be specific to the region of interest. We follow our studies by using the Local Alignment Search Tool (BLAST) in order to check for sequence specificity. Each sequence was run through Nucleotide BLAST and searched versus the human genomic for highly similar sequences [34]. Sequences which returned multiple matches in any assembly searched by the query were discarded as potential candidates for further study due to the possibility of generating reagents that may have off-target effects. After the list was narrowed down, Applied Biosystems (ABI) was selected to create custom primers and probes for each lncRNA. All remaining candidates were then run through ABI's software available through their Custom TaqMan® Assay Design Tool. ABI detected that thirteen sequences were adequate for primer/probe specific design. More details of this part of our work has been described in Dr. Barton's work [34]. In summary after careful consideration of all 42 lncRNAs differentially expressed between parous and nulliparous, we determined that lnc-BHLHE41 and lnc-EPCAM were top of the list candidates for further study. The data clearly has shown that lncBHLHE41 behaves as a tumor suppressor gene and can be responsible for the preventive effect of pregnancy in due that is highly expressed in this group of women, whereas the lncEPCAM/BC200 that is downregulated in parous and upregulated in nulliparous breast behaves as an oncogene [34].

8.4.2 The Estrogen Receptor Pathway in the Menopausal Breast

We did not observe differential expression in estrogen receptor between parous and nulliparous breasts [19]; several genes that are directly or indirectly regulated by estrogen receptor were up- or downregulated in the parous breast and were found to be enriched in the breast cancer estrogen signaling gene set [19, 20]. Among them, GATA3, an important component of this gene set, is crucial to mammary gland morphogenesis and differentiation of progenitor cells. GATA3 has been suggested to be a tumor suppressor [35], a fact supported by the observations that induction of its expression in GATA3-negative undifferentiated carcinoma cells is sufficient to induce tumor differentiation and inhibition of tumor dissemination [36–38]. The downregulation of RASD1 (RAS, dexamethasone-induced 1), a potential miR-375 target that negatively regulates ER alpha expression in breast cancer further, confirms that the genes involved in the estrogen receptor regulated pathways could be under permanent transcriptional modification as a manifestation of a higher degree of cell differentiation of the parous breast, in spite of the lack of transcriptomic differences in the levels of the receptor between parous and nulliparous breast tissues [19, 20].

8.4.3 Cell Communication in the Menopausal Breast

Cell communication, which is a key element in the process of cell and organ differentiation, is well represented in the breasts of parous women. The parous breast exhibits upregulation of desmocollin (DSC3), a calcium-dependent glycoprotein that is a member of the desmocollin subfamily of the cadherin superfamily. Members of this desmosomal family, along with the desmogleins, are found primarily in epithelial cells where they constitute the adhesive proteins of the desmosome cell-cell junction and are required for cell adhesion and desmosome formation. In addition, the upregulation of matrix Gla proteins (MGP), laminins (LAMA3 and LAMC2), and keratin 5 (KRT5) in the parous breast reflected the greater differentiated state of the breast epithelial cells [39]. This concept is supported by the observation that the loss of Matrix Gla protein expression may be associated with tumor progression and metastasis [39].

8.4.4 *Downregulated Genes in the Menopausal Breast*

Our findings that insulin-like growth factor 1 (IGF-1) is downregulated in the parous breast [19, 20] are consistent with published data reporting overall lower levels of IGF-1 in parous than in nulliparous women [40] and support the association of IGF-1 with increased breast cancer risk [40, 41]. It is known that IGF-1 stimulates mitosis and inhibits apoptosis, playing a significant role in signaling pathways involved in the pathogenesis of breast cancer. The downregulation of IGF-1 in the parous breast, in association with the significant downregulation of SOX6, EBF1 (early B-cell factor 1), ABHD5, RASD1, a potential miR-375 target that negatively regulates ER alpha expression in breast cancer [42], and RALGAPA2, could crepresent a significant driving force in the reduction of breast cancer risk conferred by pregnancy.

8.5 Relevance of the Genomic Signature in the Pre- and Postmenopausal Breast

The study of the transcriptomic profile in the premenopausal breast allowed to understand the final genomic signature of pregnancy found in the menopausal breast. For example, the observation that the "long-term constant" genes were mainly involved in cell differentiation and developmental processes (Fig. 8.3) is consistent with the transcriptomic profile described in the parous postmenopausal breast, in which upregulated genes showed enrichment of similar processes [17, 19, 20]. These findings indicate that the parity signature set after pregnancy remains until the postmenopausal years. Other development-related genes upregulated in the P breast and confirmed by RT-PCR were RASGRP1 (RAS guanyl-releasing protein 1—calcium and DAG-regulated), EGR3 (early growth response 3), and BHLHE22 (basic helix-loop-helix family member e22). These genes, in addition to differentiation, are known to regulate proliferation and cell growth [43–46].

Among the genes classified as developmental, we also found components of the WNT pathway. WNT canonical and non-canonical pathways participate in cell fate determination, cell polarity, adhesion, and motility [47, 48], all important functions in the differentiation of the breast. Differentiation induced by parity has been demonstrated to alter WNT/Notch signaling in mice [49], and we will describe in the next section of this chapter alterations in the methylation profile of genes belonging to this pathway and its regulation in the postmenopausal breast [42]. In the premenopausal breast we observed two genes of this pathway upregulated in the P breast, FZD8 (frizzled family receptor 8), which is a transmembrane receptor transducing WNT signals, and EAF2 (ELL-associated factor 2), which is important for early embryonic development and critical for adult tissue homeostasis and prevention against tumor initiation [50, 51]. WNT inhibitory factor 1, WIF1, was downregulated in P women, and although methylation of WIF1 has been observed in several tumors [52], including breast cancer [53], this can be an indication that the WNT pathway has an important role in the shifting of the stem cells to a more differentiated status in the P breast, as demonstrated earlier [18, 54]. Yet related to WNT pathway, we found FOXQ1 or forkhead box Q1 constantly downregulated in the P breast. This gene is a direct target of the canonical WNT pathway, and its overexpression has been associated with different tumors and cancer cell lines [55]. Suppression of FOXQ1 inhibits cell proliferation, motility/invasion, and epithelial-mesenchymal transition phenotypes in cancer cells [56–58]; a similar effect could be predicted in the breast of P women.

Also consistent with the transcriptomic profile of the postmenopausal breast, we observed several genes with roles in chromatin remodeling [18]. In the premenopausal breast, there were four long-term upregulated genes involved in chromatin remodeling: APOBEC3G, TOX, UHRF1, and NAP1L2. These genes interact with chromatin, either by binding with histones or recruiting histone modifiers, influencing cell proliferation and differentiation among other biological processes [59–64]. In addition, APOBEC3G (apolipoprotein B mRNA-editing enzyme, catalytic polypeptide-like 3G) is involved in RNA editing [65], and deletion in APOBEC3 gene has been correlated to breast cancer risk [66]. The upregulation of these genes in the breast of P women years after delivery indicates chromatin remodeling is enabling a permanent differentiation of the breast epithelial cells.

We also evaluated whether selected genes found in the premenopausal breast are modified in breast cancer cases. Of the genes evaluated, BHLHE22 was the most commonly modified in breast cancer (6.9% of the cases). The other genes were altered in a small percentage of breast cancer cases. Of interest is that patients who have some alteration (amplification, deep deletion, or missense mutation) in the tested genes have a lower overall survival (Fig. 8.5). Although this analysis can indicate that these genes are associated with breast cancer, our interpretation is limited because the parity status of these breast cancer cases is not available.

An important finding in the transcriptomic profile of the premenopausal is the large number of genes known to participate in immune response. The immune system has several roles in the mammary gland, being important not only for protection against pathogens, but also it is secreted into the milk and participates at the different stages of the gland development, including involution [67–70]. Of great interest, most of these immune-related genes were only upregulated in the biopsies collected within 5 years since pregnancy. This observation is consistent with our data in the postmenopausal women, in which we did not observe enrichment of immune response in an older population [17–20]. It is also in agreement with previous studies on molecular profiling of pregnancy performed in younger populations that reported changes in immune-response genes [71, 72].

Previous studies have showed differences in expression patterns of immune-related genes at distinct mammary developmental stages, before and/or after pregnancy in humans [71] and rodents [73, 74]. Rodent studies demonstrated that in the first days of mammary gland involution, there is activation of genes related to acute-phase response and inflammation [73, 74], followed by activation of monocyte and lymphoid chemokines and immunoglobulin genes [73]. The inflammatory-like environment observed during the breast involution has been proposed as one of the mechanisms that could explain the transient increase in breast cancer risk observed after each pregnancy [69, 75]. In this study, we observed that most of the genes that underwent transient changes in expression enriched biological processes related to activation and development of lymphocytes, mainly T cells [18]. Only one GO term related to inflammation (GO:0006925: inflammatory cell apoptotic process) was enriched among the transient genes [18]. This may be due to the fact that biopsies were collected at least 1 year after FTP and/or breastfeeding; thus, we may have missed an early inflammation stage. Both human and murine postpartum mammary glands have an organized influx of immune cells; however, these cells are not observed after 1 year postpartum in human, or 12 weeks in murine [69]. We have performed immunohistochemistry reactions for dendritic cells and T cell activation markers on a subset of our samples. Using antibodies anti-LAMP3 and anti-CD123 for dendritic cells, antigen presenters, we detected few positive cells and no differences between P and NP samples. When using anti-CD2 and anti-CD3D, markers of T cell activation, we observed an increase in cells positive for CD3D in P breast. This data suggest that there are differences in the number of activated T cells between P and NP breast (Fig. 8.6).

Normal pregnancy is characterized by an early expansion of regulatory T cells [76], which modulate immune tolerance during pregnancy [76]. In addition, micro chimeric cells of fetal origin that persist in the maternal circulation after delivery are postulated to provide immune surveillance with the contribution of T cells, protecting against breast cancer [77–79]. Evidence shows that micro chimeric cells are more frequent in healthy women than in breast cancer-affected women, and breast cancers without circulating micro chimeric cells are more aggressive [77–82]. In the transcriptomic profile of the premenopausal breast, we found several upregulated genes related to activation of T cells after FTP [18]. This suggests that the activation of T cells in the breast tissue of P women could trigger an early response against transformed cells, destroying them and protecting the mammary gland from neoplastic transformation. Tumor surveillance by the immune system and its impact on disease outcomes in cancer patients, and in breast cancer patients in particular, have been documented [83–86]. Finak et al. studied the stromal gene expression and clinical outcomes in breast cancer and observed that the gene set expressed predominantly in the good-outcome cluster was enriched by elements of the T helper type 1 [83]. Eight (CD2, CD8A, XCL1, CD3D, GZMA, CD247, CD48, CD52) of the genes present in this cluster [83] were also upregulated in the parous women. Poor-outcome cluster [83] showed downregulation of CXCL14, which was upregulated in the parous breast, and upregulation of CXCL1 and EDN1, which were downregulated in the parous breast. This overlap between Finak et al.'s study and ours indicates that the activation of the T cell response is a beneficial mechanism against transformed cells. Winslow et al. described gene sets with better prognostics for triple-negative breast cancers [73]. These gene sets involved cytotoxic immune response, including the genes mentioned above, and HLA encoding genes [84]. HLA-DRA (major histocompatibility complex, class II, DR alpha) and HLA-DPA1 (major histocompatibility complex, class II, DP alpha 1) were constantly activated in our study. The upregulation of HLA-DPA1 has been associated with a benign behavior of certain neurological tumors and related to the immune defense-associated genes [87]. Therefore, a similar role could be attributed to the upregulation of these two genes in the cancer preventative effect of pregnancy in the premenopausal women.

Altogether, parous premenopausal breast shows a specific transcriptome profile, in which genes controlling chromatin remodeling and cell differentiation are activated after FTP and stay upregulated for many years and are the hallmark of the postmenopausal breast [17–19]. Of significant importance is that the genes involved in immune response were in majority related to T cell activation, and these were activated soon after the FTP; however, their expression returns to levels similar to those observed in the nulliparous breast 5 or more years after FTP. These transcripts may work in protecting the mammary gland against neoplastically transformed cells through T cells. However, because this immune surveillance appears transitory, we infer that cell differentiation, activated by the genes whose expression was permanently affected by parity, may be the main molecular mechanism responsible for the preventive effect of parity against breast cancer.

The demonstration that differentiation of the breast induced by an early pregnancy imprints a specific genomic signature [17–20] that can be detected in postmenopausal women has allowed us to identify enriched biological processes and pathways such as RNA related processes, differentiation, and development of epidermis and ectoderm, and cell-substrate junction assembly, whereas in the case of downregulated genes, the biological processes that were enriched included IGF-like growth factor signaling, somatic stem cell maintenance, and apoptosis. Pathways that were enriched by upregulated genes included breast cancer estrogen signaling, cell communication, and mRNA processing machinery. Numerous pathways were enriched by downregulated genes; the most significant ones were the insulin, Wnt, and integrins signaling pathways, MAPK, cytokine-cytokine receptor interaction, tight junction, and focal adhesion, all representing proteins that are highly expressed in malignancies.

A very novel contribution of our studies is that we were able to demonstrate that the genomic signature suggests that the differentiation process of breast cells is centered in the mRNA processing reactome, which emerges as an important regulatory pathway induced by pregnancy [19, 20]. The biological importance of the differential expression of genes that control the spliceosome could be an indication of a safeguard mechanism at post-transcriptional level that maintains the fidelity of the transcriptional process. In addition, the critical regulatory pre-mRNA splicing mechanism could also regulate the expression of specific genes controlling estrogen signaling pathways, cell communication, and differentiation, as well as pathways related to chromatin remodeling, altogether resulting in control of cell differentiation and breast cancer prevention [19, 20].

8.6 Role of Epigenetic Changes in the Signature of Prevention

Our observations that during the postmenopausal years the breast of both parous and nulliparous women contains preponderantly Lob 1, and the fact that nulliparous women are at higher risk of developing breast cancer than parous women, indicate that Lob 1 in these two groups of women differ biologically and exhibit different susceptibility to carcinogenesis [8].

The breast tissues of the P and NP women contained ducts and Lob 1 [15, 16]. The microscopic analysis of the breast tissue revealed that the population of luminal cells lining ducts and Lob 1 was composed of cells that were characterized by their nuclear appearance into two types: one that contained large and palely stained nuclei with prominent nucleoli and another consisting of small hyper chromatic nuclei [19]. The pale staining of the large former nuclei is a feature indicative of a high content of non-condensed euchromatin; these nuclei were called euchromatin-rich nuclei (EUN). The hyperchromasia observed in the latter nuclei was indicative of chromatin condensation and high content of heterochromatin; these nuclei were identified as heterochromatin-rich nucleus (HTN). The analysis of the distribution of HTN and EUN cells in histological sections of the breast core biopsies revealed that EUN were more abundant in the NP than in the P breast tissues, whereas the inverse was true for the HTN; these differences were statistically significant [19]. These observations indicated that a shift of the EUN cell population to a more densely packed chromatin cell (HTN) had occurred in association with the history of pregnancy as a distinctive pattern of the postmenopausal parous breast [19] (Fig. 8.10).

Since chromatin condensation is part of the process of chromatin remodeling toward gene silencing that is highly regulated by methylation of histones, we verified this phenomenon by immunohistochemistry (IHC) incubating NP and P breast tissues with antibodies against histone 3 dimethylated at lysine 9 (H3K9me2) and trimethylated at lysine 27 (H3K27me3) [18] (Fig. 8.10). The IHC stain revealed that methylation of H3 at both lysine 9 and 27 was increased in the heterochromatin condensed nuclei of epithelial cells of the parous breast when compared to the euchromatin-rich nuclei of the nulliparous breast. In the nulliparous breast, the reactivity in individual cells was less intense, and the number of positive cells was significantly lower. These variations in chromatin reorganization were supported by the upregulation of CBX3, CHD2, L3MBTL, and EZH2 genes controlling this process [19] (Fig. 8.10).

Our data clearly indicate that there are morphological indications of chromatin remodeling in the parous breast, such as the increase in the number of epithelial cells with condensed chromatin and increased reactivity with anti-H3K9me2 and H3K27me3 antibodies. Histone methylation is a major determinant for the formation of active and inactive regions of the genome and is crucial for the proper programming of the genome during development [34, 88]. In the parous breast, there is upregulation of transcription factors and chromatin remodeling genes such as CHD2 or chromodomain helicase DNA-binding protein 2 and the CBX3 or Chromobox homolog 3, whose products are required for controlling recruitment of protein/protein or DNA/protein interactions.

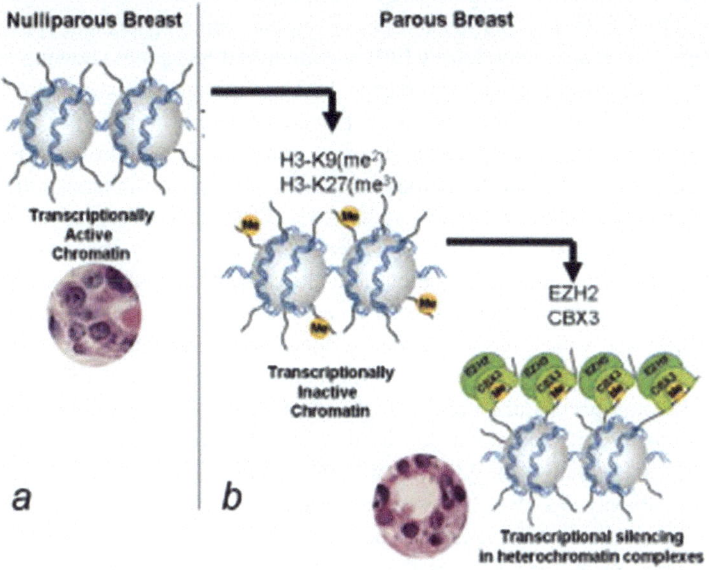

Fig. 8.10 (**a**) Transcriptionally active chromatin is predominantly expressed in the euchromatin-rich nuclei (EUN) of the nulliparous woman's breast. (**b**) Transcriptionally inactive chromatin is more frequently found in the heterochromatin-rich nuclei (HTN) of the parous breast; its presence is associated with histone 3 methylation at lysines 9 and 27 and transcriptional silencing in heterochromatin complexes [18]

CBX3 is involved in transcriptional silencing in heterochromatin-like complexes and recognizes and binds H3 tails methylated at lysine 9, leading to epigenetic repression. Two other important genes related to the polycomb group (PcG) protein that are upregulated in the parous breast are the L3MBTL gene or l(3)mbt-like and the histone-lysine N-methyl transferase, or EZH2. Members of the PcG form multimeric protein complexes that maintain the transcriptional repressive state of genes over successive cell generations. EZH2 is an enzyme that acts mainly as a gene silencer, performing this role by the addition of three methyl groups to lysine 27 of histone 3, a modification that leads to chromatin condensation [88–90].

8.6.1 *Methylation Changes in the DNA of Parous Women Are Part of Chromatin Remodeling and the Genomic Signature of Pregnancy*

The chromatin remodeling process is demonstrated not only by the shifting of the EUN to the HTN cells, but also confirmed by the increase in methylation of histones H3K9me2 and H3K27me3. This is an indication that methylation of other genes could also be involved in the process. For this purpose we went one step further, and using the DNA from nulliparous and parous breast core biopsies and applying the MBD-cap sequencing methodology [90], we identified 583 genes showing different levels of methylation between the parous and nulliparous breasts. From the 583 genes, 455 were hyper methylated in the parous while 128 were hyper methylated in the nulliparous breast, confirming the reprogramming of the chromatin to a more silenced or resting stage in the parous breast. To get a better understanding of the methylation profile of the 583 genes, we used a software called Integrative Genomics Viewer (IGV) [68, 91–93]. The IGV allowed us to identify the distinct areas, throughout the entire gene, where the methylation levels differed between the sample groups. The identification of these areas, known as differentially methylated regions (DMRs), was important because they were more likely to affect gene expression as has been shown in the literature. We performed the comparison between the nulliparous and parous methylation profiles against the human reference genome "hg 18" and against each other [68]. We found that, for example, the gene COBRA 1, which is the cofactor of BRCA1 and has been shown to work in its regulatory pathway [94], was hyper methylated in the nulliparous breast. The methylation levels for each sample at each base pair showed an area of higher methylation occurring in at least four of the samples of one group as compared to all members of the opposing group; that area was defined as a (DMR). COBRA 1 had a DMR near the end of the gene. Every gene within the 583 gene list was closely examined for DMRs, and the chromosomal locations at which these DMRs were found and marked were published by us [42].

After analysis of the 583 genes using the IGV, we have identified the DMRs of 53 genes. Of the 455 parous hyper methylated genes, 41 had DMRs. The exact locations of these DMRs were published in the journal *Genes* in 2014 [42]. Of the 128 nulliparous hyper methylated genes, 12 had DMRs [42]. Analysis and research into the functions of these 53 genes identified seven which interacted with each other in either the Wnt signaling pathway or its controlling PI3K/AKT/mTOR pathways. An overview of the involvement in the canonical Wnt pathway is shown in Fig. 8.11. The interworking of these genes with each other and with other genes within the statistically methylated 583 can be seen in Fig. 8.12 [42].

Of the seven genes with DMRs which we have shown to work together in the Wnt pathway or its controllers, three worked directly in canonical Wnt signaling. Interestingly, when we analyzed the genes differentially expressed between parous and nulliparous [18, 42], we found genes that also participate in the Wnt pathway, such as CSNK1A1 and SOX family. FZD1, which is hyper methylated in the nulliparous breast, codes for the Frizzled receptor. When activated, this receptor directly activates Disheveled (Dsh) in the cytosol to begin the Wnt signaling cascade [95]. GSK3B, which also contains DMRs hyper methylated in the nulliparous women, has as main rule to decrease beta-catenin levels in the Wnt signaling pathway [96]. PPP2CA (PP2A) is suggested to work both upstream and downstream of beta-catenin to assist in its stabilization [60, 61]. DACT1 assists in Wnt signaling by upregulating GSK3B [97, 98]. ROBO1, INPP4B, and IL6ST genes are active in PI3K-dependent AKT signaling [99, 100].

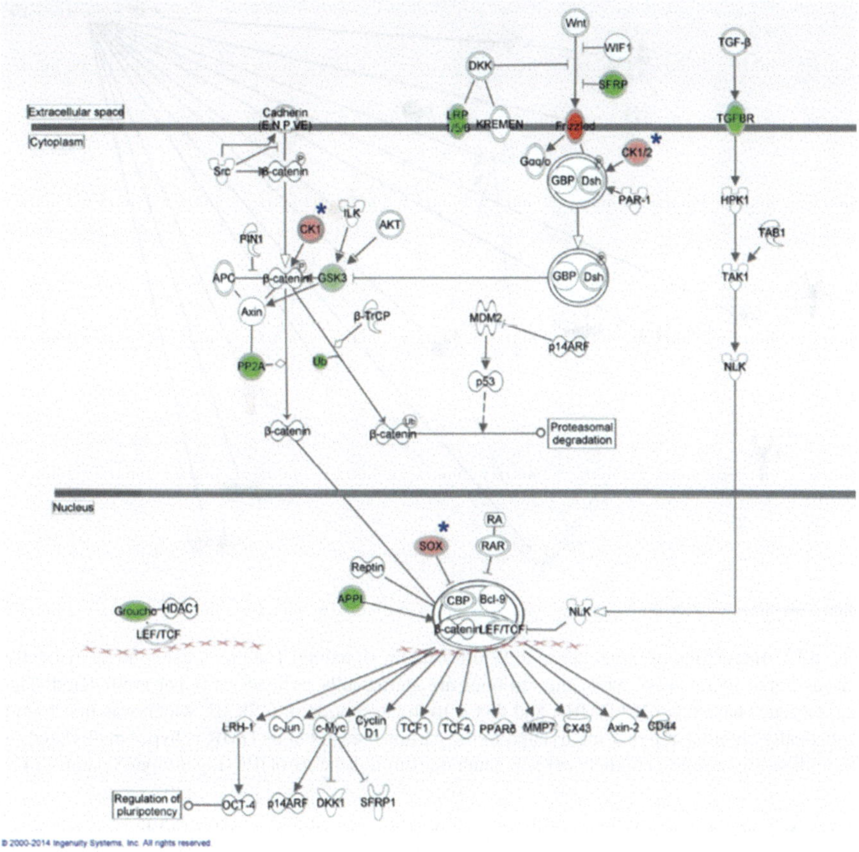

Fig. 8.11 Canonical WNT/β-catenin signaling genes marked in green are hyper methylated in parous women (suggesting downregulation of the gene in parous women). Genes in red are hyper methylated within nulliparous women. Genes marked with an asterisk (*) were observed differentially expressed the microarray data. This canonical pathway was generated through the use of IPA (Ingenuity® Systems) [42]

The potential significance of the Wnt signaling pathway is rooted in an experiment performed in 1982 to find which genes would mutate in mice injected with mouse mammary tumor virus locating int1, a proto-oncogene [101, 102]. Int1 was soon found to be highly conserved across multiple species, including drosophila and humans. Int1 was discovered to be the mammalian homologue of the drosophila Wingless (Wg), a gene previously found to be a segment polarity gene in embryonic development. The Wnt signaling pathway was given its name from the combination of Wg and int1 and has always had a close relationship to both differentiation and breast cancer. We have discussed the potential role of Wnt in the preventive effect of pregnancy in our previous publication [42], and I will not go into detail here but

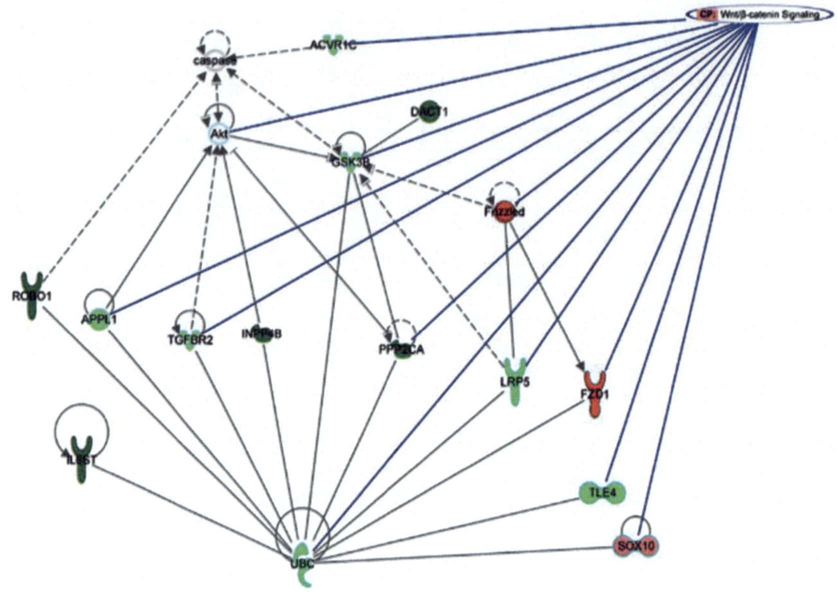

Fig. 8.12 Interaction of target genes in Wnt/β-catenin signaling. The green genes are statistically parous hyper methylated, while the red ones are statistically nulliparous hyper methylated. The darker genes have recorded DMRs, and this is to the exception of GSK3B, which was first found statistically significant hyper methylated in the parous breast, but its DMR is hyper methylated in the nulliparous samples. This network was generated through the use of IPA (Ingenuity®Systems) [42]

only indicate that the Wnt signaling pathway is significantly affected by the preventive agent human chorionic gonadotropin.

8.6.2 *Relevance of Gene Methylation in Breast Cancer Prevention*

Our work [17, 19, 20, 42] has clearly demonstrated that the breast of parous postmenopausal women exhibits a specific signature that has been induced by a full-term pregnancy. This signature reveals for the first time that the differentiation process is centered in chromatin remodeling by epigenetic changes induced by methylation of specific genes that are important regulatory pathways induced by pregnancy.

8.7 Manipulating Chromatin Remodeling

In the postmenopausal years the breast of both parous and nulliparous women contains preponderantly morphologically similar lobules type 1; they differ in exhibiting different susceptibility to developing cancer, and in their genomic profile which includes upregulation of long noncoding RNAs such as XIST and chromatin remodeling genes, as well as in the presence of histone H3 methylation on lysine 27 (H3K27me3) and 9 (H3K9me2) in epithelial cells [19, 42]. Mimicking pregnancy by using r-hCG in rats induces the differentiation of the mammary gland and gains the same protection against developing mammary tumors [13]. hCG may play an important role in the protection induced by pregnancy. In supporting this concept, Toniolo et al. reported that maternal breast cancer risk decreased with increasing hCG levels especially for pregnancies before 25 years of age [103]. Furthermore, women with a maximum body mass index (BMI) < 27.5 exhibited a significant protective effect against breast cancer after r-hCG use [104], suggesting r-hCG is able to remodel mammary epithelial cells, leading to a new cell state that is resistant to malignancy.

With Dr. Lee K. Tay in the early 1980s [12, 105] we have shown that the parous rat has a more efficient DNA repair capacity in mammary epithelial cells, suggesting the difference in DNA damage response is related to the parity induced protection. In supporting of this data, p53, the tumor suppressor gene that has many mechanisms of anticancer, was observed in 1997 by Dr. Pramod Srivastava and Dr. Irma Russo that was upregulated in parous and r-hCG-treated rats [106]. It was further demonstrated that p53 is required for hormone-mediated protection of mammary tumorigenesis [107]. Interestingly, the analysis of transcriptome of the parous and nulliparous postmenopausal breast showed the changes of long noncoding RNA and chromatin remodeling genes. XIST, the long noncoding RNA gene responsible for the initiation of X chromosome inactivation, is upregulated in parous breast [19].

Based on this previous research, we embarked on further exploring the mechanism underlying the protection induced by r-hCG that in turn would help us to reveal how early FTP has an important consequence on breast cancer protection later in life, and provide the basis for the use of r-hCG as a preventive strategy for breast cancer. r-hCG treated MCF-10F cells increased XIST expression. Consistent with the upregulation of XIST, which is important for the inactivation of X chromosome, HDAC8, an X-linked gene, is downregulated in the parous breast [17, 19, 20]. Inhibition of HDAC8 remarkably enhances p53 acetylation and p53 protein in leukemia stem cells [108, 109]. HDAC6, another X-linked gene, also has similar effect on regulating p53; HDAC6 inhibition differentially modulates p53 by upregulating wild-type p53 [110]. Therefore, we concluded that upregulation of XIST by r-hCG regulates p53 through epigenetic regulation of X-linked regulatory genes. These findings were supported by data in the literature indicating that X-linked genes differentially expressed between mammary tumors of mouse model and normal mammary glands are predominantly related to chromatin remodeling, chromosome segregation, and translational control [111].

Loss of X chromosome inactivation is a common phenomenon in breast cancers [112], and there are publications supporting this idea; X-linked genes HUWE1 [113], RbAp46 [114], and BRCC36 [115] are shown to be involved in BRCA1 regulation. BRCA1 is the best-known breast cancer susceptibility protein linked to breast cancer risk, and one of its functions is to repair cell damage. Although the association between XIST and BRCA1 is controversial, we postulated with Yanrong Su that not only BRCA1 can regulate XIST, but also XIST regulates BRCA1 through epigenetic control of regulatory genes on X chromosome. In addition, increased global H3K27me3 and H3K9me2 modifications were consistently observed in parous breast epithelium and r-hCG-treated MCF-10F cells. Histone lysine methylation is a major determinant for the formation of active or inactive chromatin and is crucial for the proper programming of the genome during development [116]. The chromatin remodeling genes such as EZH2, CHD2, CBX3, AEBP2, and L3MBTL were upregulated in the parous breast [19]. EZH2 is a key regulator in controlling cellular differentiation. EZH2 catalyzes the tri-methylation of histone H3 on Lys 27 (H3K27me3) to generate epigenetic mark [117]. The study that explored H3K27me3 marks in different breast cancer subtypes showed H3K27me3 is inversely associated with tumor grade, and is the least abundant in triple-negative and HER2 tumors [118]. Consistently, Yan et al. reported that H3K27me3 activation by inhibiting demethylases JMJD3 and UTX leads to inhibition of breast cancer stem cells and self-renewal, as well as global level of stemness factors [119]. These observations lead us to postulate that the histone modification is also involved in the protection induced by pregnancy and r-hCG by regulating a subset of genes.

The relevance of this line of research is based from the fact that r-hCG mimics pregnancy. The mechanism of this protection is that long noncoding RNA XIST is upregulated in parous breast and in r-hCG-treated MCF-10F cells and that the X-linked genes are associated with the regulation of breast and ovarian cancer susceptibility protein BRCA1 and p53. The hCG receptor is predominantly found in the breast and ovary, and early full-term pregnancy decreases the risk of developing breast cancer and ovarian cancer; and lastly increased histone H3 methylation on lysine 27 and lysine 9 is consistently observed in the parous breast and r-hCG-treated MCF-10F cells. All of these lead us to postulate that increased XIST and H3K27me3 induced by parity and r-hCG treatment regulate the expression and function of genes (especially BRCA1 and p53) involved in cell differentiation, DNA repair, and prevention of neoplastic transformation, thus providing protection against breast cancer.

This work has clinical translation implications; given the fact that modern society is changing our lives, more and more women choose to have their first child later in life, and at that time the environment or other exposures in their early lives may have already caused the changes in the mammary gland genome and abrogated the protection offered by pregnancy. The understanding of how early pregnancy and hCG induce protection provides the possibility to use r-hCG in early life as a preventative strategy against breast cancer. In addition, the panel of genes differentially expressed by hCG treatment may have the potential to be biomarkers to predict the protection against breast cancer. Moreover, for those women at high risk, such as women with BRCA1/BRCA2 mutation and women with strong family history, if we demonstrate in our ongoing clinical trial (https://clinicaltrials.gov/beta/with the Clinical trials Gov identifier:NGT00700778) that r-hCG is able to induce protection against breast cancer through epigenetic control of p53, BRCA1, and other genes, it would be greatly helpful as nowadays the only way to reduce their risk is bilateral prophylactic mastectomy.

References Cited

1. Breasted, J. H. (Ed.). (1991). *The Edwin Smith Surgical Papyrus: published in facsimile and hieroglyphic transliteration with translation and commentary in two volumes* (Vol. 1). Chicago: University of Chicago Press.
2. Mustacchi, P. (1961). Ramazzini and Rigoni-Stern on parity and breast cancer. Clinical impression and statistical corroboration. *Archives of Internal Medicine, 108*, 639–642.
3. MacMahon, B., Cole, P., Lin, T. M., Lowe, C. R., Mirra, A. P., Ravnihar, B., Salber, E. J., Valaoras, V. G., & Yuasa, S. (1970). Age at first birth and breast cancer risk. *Bulletin of the World Health Organization, 43*, 209–221.
4. Russo, I. H., & Russo, J. (1978). Developmental stage of the rat mammary gland as determinant of its susceptibility to 7, 12-dimethylbenz (a) anthracene. *Journal of the National Cancer Institute, 61*, 1439–1449.
5. Russo, J., & Russo, I. H. (1978). DNA labeling index and structure of the rat mammary gland as determinant of its susceptibility to carcinogenesis. *Journal of the National Cancer Institute, 61*, 1451–1459.
6. Clarke, C. A., Purdie, D. M., & Glaser, S. L. (2006). Population attributable risk of breast cancer in white women associated with immediately modifiable risk factors. *BMC Cancer, 6*, 170.
7. Jemal, A., Siegel, R., Ward, E., Murray, T., Xu, J., & Thun, M. J. (2007). Cancer statistics, 2007. *CA: a Cancer Journal for Clinicians, 57*, 43–66.
8. Russo, J., Balogh, G. A., & Russo, I. H. (2008). Full-term pregnancy induces a specific genomic signature in the human breast. *Cancer Epidemiology, Biomarkers & Prevention, 17*, 51–66.
9. Thordarson, G., Jin, E., Guzman, R. C., Swanson, S. M., Nandi, S., & Talamantes, F. (1995). Refractoriness to mammary tumorigenesis in parous rats: Is it caused by persistent changes in the hormonal environment or permanent biochemical alterations in the mammary epithelia? *Carcinogenesis, 16*, 2847–2853.
10. Sinha, D. K., Pazik, J. E., & Dao, T. L. (1988). Prevention of mammary carcinogenesis in rats by pregnancy: Effect of full-term and interrupted pregnancy. *British Journal of Cancer, 57*, 390–394.
11. Russo, J., & Russo, I. H. (1980). Influence of differentiation and cell kinetics on the susceptibility of the rat mammary gland to carcinogenesis. *Cancer Research, 40*, 2677–2687.
12. Tay, L. K., & Russo, J. (1981). Formation and removal of 7, 12-dimethylbenz[a]anthracene–nucleic acid adducts in rat mammary epithelial cells with different susceptibility to carcinogenesis. *Carcinogenesis, 2*, 1327–1333.
13. Russo, I. H., Koszalka, M., & Russo, J. (1991). Comparative study of the influence of pregnancy and hormonal treatment on mammary carcinogenesis. *British Journal of Cancer, 64*, 481–484.
14. Fisher, D. A. (2006). Fetal and neonatal endocrinology. In D. G. LJ & J. L. Jameson (Eds.), *Endocrinology*. Philadelphia: Elsevier Saunders.
15. Russo, J., Moral, R., Balogh, G. A., Mailo, D., & Russo, I. H. (2005). The protective role of pregnancy in breast cancer. *Breast Cancer Research, 7*, 131–142.

16. Russo, J., & Russo, I. H. (1997). Role of differentiation in the pathogenesis and prevention of breast cancer. *Endocrine-Related Cancer, 4*, 7–21.
17. Belitskaya-Levy, I., Zeleniuch-Jacquotte, A., Russo, J., Russo, I. H., Bordas, P., Ahman, J., Afanasyeva, Y., Johansson, R., Lenner, P., Li, X., de Cicco, R. L., Peri, S., Ross, E., Russo, P. A., Santucci-Pereira, J., Sheriff, F. S., Slifker, M., Hallmans, G., Toniolo, P., & Arslan, A. A. (2011). Characterization of a genomic signature of pregnancy identified in the breast. *Cancer Prevention Research, 4*, 1457–1464.
18. Santucci-Pereira, J., Zeleniuch-Jacquotte, A., Afanasyeva, Y., Zhong, H., Slifker, M., Peri, S., Ross, E. A., López de Cicco, R., Zhai, Y., Nguyen, T., Sheriff, F., Russo, I. H., Su, Y., Arslan, A. A., Bordas, P., Lenner, P., Åhman, J., Landström Eriksson, A. S., Johansson, R., Hallmans, G., Toniolo, P., & Russo, J. (2019). Genomic signature of parity in the breast of premenopausal women. *Breast Cancer Research, 21*, 46. https://doi.org/10.1186/s13058-019-1128-x. PMID:30922380
19. Russo, J., Santucci-Pereira, J., de Cicco, R. L., Sheriff, F., Russo, P. A., Peri, S., Slifker, M., Ross, E., Mello, M. L., Vidal, B. C., Belitskaya-Levy, I., Arslan, A., Zeleniuch-Jacquotte, A., Bordas, P., Lenner, P., Ahman, J., Afanasyeva, Y., Hallmans, G., Toniolo, P., & Russo, I. H. (2012). Pregnancy-induced chromatin remodeling in the breast of postmenopausal women. *International Journal of Cancer, 131*, 1059–1070.
20. Peri, S., de Cicco, R. L., Santucci-Pereira, J., Slifker, M., Ross, E. A., Russo, I. H., Russo, P. A., Arslan, A. A., Belitskaya-Levy, I., Zeleniuch-Jacquotte, A., et al. (2012). Defining the genomic signature of the parous breast. *BMC Medical Genomics, 5*, 46–57.
21. Herrmann, A., Fleischer, K., Czajkowska, H., Müller-Newen, G., & Becker, W. (2007). Characterization of cyclin L1 as an immobile component of the splicing factor compartment. *The FASEB Journal, 21*, 3142–3152.
22. Santucci-Pereira, J., Weng, S., Slifker, M., Russo, J. (2015). RNA splicing events are related to breast cancer prevention. *Proceedings of the American Association for Cancer Research, 56*, 1973a.
23. Loyer, P., Trembley, J. H., Grenet, J. A., Busson, A., Corlu, A., Zhao, W., Kocak, M., Kidd, V. J., & Lahti, J. M. (2008). Characterization of cyclin L1 and L2 interactions with CDK11 and splicing factors: Influence of cyclin L isoforms on splice site selection. *The Journal of Biological Chemistry, 283*, 7721–7732.
24. Li, H. L., Wang, T. S., Li, X. Y., Li, N., Huang, D. Z., Chen, Q., & Ba, Y. (2007). Overexpression of cyclin L2 induces apoptosis and cell-cycle arrest in human lung cancer cells. *Chinese Medical Journal (Engl), 120*, 905–909.
25. Zhuo, L., Gong, J., Yang, R., Sheng, Y., Zhou, L., Kong, X., & Cao, K. (2009). Inhibition of proliferation and differentiation and promotion of apoptosis by cyclin L2 in mouse embryonic carcinoma P19 cells. *Biochemical and Biophysical Research Communications, 390*, 451–457.
26. Erwin, J. A., & Lee, J. T. (2010). Characterization of X-chromosome inactivation status in human pluripotent stem cells. *Current Protocols in Stem Cell Biology*. Chapter 1: Unit 1B 6.
27. Vincent-Salomon, A., Ganem-Elbaz, C., Manié, E., Raynal, V., Sastre-Garau, X., Stoppa-Lyonnet, D., Stern, M. H., & Heard, E. (2007). X inactive-specific transcript RNA coating and genetic instability of the X chromosome in BRCA1 breast tumors. *Cancer Research, 67*, 5134–5140.
28. Xiao, C., Sharp, J. A., Kawahara, M., Davalos, A. R., Difilippantonio, M. J., Hu, Y., Li, W., Cao, L., Buetow, K., Ried, T., Chadwick, B. P., Deng, C. X., & Panning, B. (2007). The XIST noncoding RNA functions independently of BRCA1 in X inactivation. *Cell, 28*, 977–989.
29. Silver, D. P., Dimitrov, S. D., Feunteun, J., Gelman, R., Drapkin, R., Lu, S. D., Shestakova, E., Velmurugan, S., Denunzio, N., Dragomir, S., Mar, J., Liu, X., Rottenberg, S., Jonkers, J., Ganesan, S., & Livingston, D. M. (2007). Further evidence for BRCA1 communication with the inactive X chromosome. *Cell, 128*, 991–1002.

30. Breton, C., Di Scala-Guenot, D., & Zingg, H. H. (2001). Oxytocin receptor gene expression in rat mammary gland: Structural characterization and regulation. *Journal of Molecular Endocrinology, 27*, 175–189.
31. Russo, I. H., & Russo, J. (2011). Pregnancy-induced changes in breast cancer risk. *Journal of Mammary Gland Biology and Neoplasia, 16*, 221–233.
32. Koshimizu, T. A., Fujiwara, Y., Sakai, N., Shibata, K., & Tsuchiya, H. (2010). Oxytocin stimulates expression of a noncoding RNA tumor marker in a human neuroblastoma cell line. *Life Sciences, 86*, 455–460.
33. Matthew, G. G., & Young, R. A. (2010). Repressive transcription. *Science, 329*, 150–151.
34. Barton, M., Santucci-Pereira, J., Vaccaro, O. G., Nguyen, T., Su, Y., & Russo, J. (2019). BC200 overexpression contributes to luminal and triple negative breast cancer pathogenesis. *BMC Cancer, 19*, 994.
35. Chou, J., Provot, S., & Werb, Z. (2010). GATA3 in development and cancer differentiation: Cells GATA have it! *Journal of Cellular Physiology, 222*, 42–49.
36. Pei, X. H., Bai, F., Smith, M. D., Usary, J., Fan, C., Pai, S. Y., Ho, I. C., Perou, C. M., & Xiong, Y. (2009). CDK inhibitor p18(INK4c) is a downstream target of GATA3 and restrains mammary luminal progenitor cell proliferation and tumorigenesis. *Cancer Cell, 15*, 389–401.
37. Kouros-Mehr, H., Bechis, S. K., Slorach, E. M., Littlepage, L. E., Egeblad, M., Ewald, A. J., Pai, S. Y., Ho, I. C., & Werb, Z. (2008). GATA-3 links tumor differentiation and dissemination in a luminal breast cancer model. *Cancer Cell, 13*, 141–152.
38. Fischer, J., Klein, P. J., Farrar, G. H., Hanisch, F. G., & Uhlenbruck, G. (1984). Isolation and chemical and immunochemical characterization of the peanut-lectin-binding glycoprotein from human milk-fat-globule membranes. *The Biochemical Journal, 224*, 581–589.
39. Chen, L., O'Bryan, J. P., Smith, H. S., & Liu, E. (1990). Overexpression of matrix Gla protein mRNA in malignant human breast cells: isolation by differential cDNA hybridization. *Oncogene, 5*, 1391–1395.
40. Holmes, M. D., Pollak, M. N., & Hankinson, S. E. (2002). Lifestyle correlates of plasma insulin-like growth factor I and insulin-like growth factor binding protein 3 concentrations. *Cancer Epidemiology, Biomarkers & Prevention, 11*, 862–867.
41. Key, T. J., Appleby, P. N., Reeves, G. K., & Roddam, A. W. (2010). Insulin-like growth factor 1 (IGF1), IGF binding protein 3 (IGFBP3), and breast cancer risk: Pooled individual data analysis of 17 prospective studies. *The Lancet Oncology, 11*, 530–542.
42. Russo, J., Santucci-Pereira, J., & Russo, I. H. (2014). The genomic signature of breast cancer prevention. *Genes, 5*, 65–83.
43. Huang, H., Jin, T., Wang, L., Wang, F., Zhang, R., Pan, Y., et al. (2013). The RAS guanyl nucleotide-releasing protein RasGRP1 is involved in lymphatic development in zebrafish. *The Journal of Biological Chemistry, 288*, 2355–2364.
44. Kortum, R. L., Rouquette-Jazdanian, A. K., & Samelson, L. E. (2013). Ras and extracellular signal-regulated kinase signaling in thymocytes and T cells. *Trends in Immunology, 34*, 259–268.
45. To SQ, Knower, K. C., & Clyne, C. D. (2013). NFkappaB and MAPK signalling pathways mediate TNFalpha-induced early growth response gene transcription leading to aromatase expression. *Biochemical and Biophysical Research Communications, 433*, 96–101.
46. Dugas, J. C., Ibrahim, A., & Barres, B. A. (2012). The T3-induced gene KLF9 regulates oligodendrocyte differentiation and myelin regeneration. *Molecular and Cellular Neurosciences, 50*, 45–57.
47. Wang, H. Q., Xu, M. L., Ma, J., Zhang, Y., & Xie, C. H. (2012). Frizzled-8 as a putative therapeutic target in human lung cancer. *Biochemical and Biophysical Research Communications, 417*, 62–66.
48. Katoh, M. (2008). WNT signaling in stem cell biology and regenerative medicine. *Current Drug Targets, 9*, 565–570.
49. Meier-Abt, F., Milani, E., Roloff, T., Brinkhaus, H., Duss, S., Meyer, D. S., et al. (2013). Parity induces differentiation and reduces Wnt/Notch signaling ratio and proliferation poten-

tial of basal stem/progenitor cells isolated from mouse mammary epithelium. *Breast Cancer Research, 15*, R36.

50. Wan, X., Ji, W., Mei, X., Zhou, J., Liu, J. X., Fang, C., et al. (2010). Negative feedback regulation of Wnt4 signaling by EAF1 and EAF2/U19. *PLoS One, 5*, e9118.
51. Su, F., Pascal, L. E., Xiao, W., & Wang, Z. (2010). Tumor suppressor U19/EAF2 regulates thrombospondin-1 expression via p53. *Oncogene, 29*, 421–431.
52. Wissmann, C., Wild, P. J., Kaiser, S., Roepcke, S., Stoehr, R., Woenckhaus, M., et al. (2003). WIF1, a component of the Wnt pathway, is down-regulated in prostate, breast, lung, and bladder cancer. *The Journal of Pathology, 201*, 204–212.
53. Ai, L., Tao, Q., Zhong, S., Fields, C. R., Kim, W. J., Lee, M. W., et al. (2006). Inactivation of Wnt inhibitory factor-1 (WIF1) expression by epigenetic silencing is a common event in breast cancer. *Carcinogenesis, 27*, 1341–1348.
54. Russo, J., Balogh, G. A., Chen, J., Fernandez, S. V., Fernbaugh, R., Heulings, R., et al. (2006). The concept of stem cell in the mammary gland and its implication in morphogenesis, cancer and prevention. *Frontiers in Bioscience, 11*, 151–172.
55. Christensen, J., Bentz, S., Sengstag, T., Shastri, V. P., & Anderle, P. (2013). FOXQ1, a novel target of the Wnt pathway and a new marker for activation of Wnt signaling in solid tumors. *PLoS One, 8*, e60051.
56. Sehrawat, A., Kim, S. H., Vogt, A., & Singh, S. V. (2013). Suppression of FOXQ1 in benzyl isothiocyanate-mediated inhibition of epithelial-mesenchymal transition in human breast cancer cells. *Carcinogenesis, 34*, 864–873.
57. Qiao, Y., Jiang, X., Lee, S. T., Karuturi, R. K., Hooi, S. C., & Yu, Q. (2011). FOXQ1 regulates epithelial-mesenchymal transition in human cancers. *Cancer Research, 71*, 3076–3086.
58. Gao, M., Shih Ie, M., & Wang, T. L. (2012). The role of forkhead box Q1 transcription factor in ovarian epithelial carcinomas. *International Journal of Molecular Sciences, 13*, 13881–13893.
59. Tessema, M., Yingling, C. M., Grimes, M. J., Thomas, C. L., Liu, Y., Leng, S., et al. (2012). Differential epigenetic regulation of TOX subfamily high mobility group box genes in lung and breast cancers. *PLoS One, 7*, e34850.
60. Aliahmad, P., de la Torre, B., & Kaye, J. (2010). Shared dependence on the DNA-binding factor TOX for the development of lymphoid tissue-inducer cell and NK cell lineages. *Nature Immunology, 11*, 945–952.
61. Aliahmad, P., Kadavallore, A., de la Torre, B., Kappes, D., & Kaye, J. (2011). TOX is required for development of the CD4 T cell lineage gene program. *Journal of Immunology, 187*, 5931–5940.
62. Jin, W., Chen, L., Chen, Y., Xu, S. G., Di, G. H., Yin, W. J., et al. (2010). UHRF1 is associated with epigenetic silencing of BRCA1 in sporadic breast cancer. *Breast Cancer Research and Treatment, 123*, 359–373.
63. Attia, M., Forster, A., Rachez, C., Freemont, P., Avner, P., & Rogner, U. C. (2011). Interaction between nucleosome assembly protein 1-like family members. *Journal of Molecular Biology, 407*, 647–660.
64. Attia, M., Rachez, C., De Pauw, A., Avner, P., & Rogner, U. C. (2007). Nap1l2 promotes histone acetylation activity during neuronal differentiation. *Molecular and Cellular Biology, 27*, 6093–6102.
65. Lackey, L., Law, E. K., Brown, W. L., & Harris, R. S. (2013). Subcellular localization of the APOBEC3 proteins during mitosis and implications for genomic DNA deamination. *Cell Cycle, 12*, 762–772.
66. Long, J., Delahanty, R. J., Li, G., Gao, Y. T., Lu, W., Cai, Q., et al. (2013). A common deletion in the APOBEC3 genes and breast cancer risk. *Journal of the National Cancer Institute, 105*, 573–579.
67. Monks, J., Geske, F. J., Lehman, L., & Fadok, V. A. (2002). Do inflammatory cells participate in mammary gland involution? *Journal of Mammary Gland Biology and Neoplasia, 7*, 163–176.

68. Csanaky, K., Doppler, W., Tamas, A., Kovacs, K., Toth, G., & Reglodi, D. (2014). Influence of terminal differentiation and PACAP on the cytokine, chemokine, and growth factor secretion of mammary epithelial cells. *Journal of Molecular Neuroscience, 52*, 28–36.
69. Martinson, H. A., Jindal, S., Durand-Rougely, C., Borges, V. F., & Schedin, P. (2015). Wound healing-like immune program facilitates postpartum mammary gland involution and tumor progression. *International Journal of Cancer, 136*, 1803–1813.
70. Plaks, V., Boldajipour, B., Linnemann, J. R., Nguyen, N. H., Kersten, K., Wolf, Y., et al. (2015). Adaptive immune regulation of mammary postnatal organogenesis. *Developmental Cell, 34*, 493–504.
71. Asztalos, S., Gann, P. H., Hayes, M. K., Nonn, L., Beam, C. A., Dai, Y., et al. (2010). Gene expression patterns in the human breast after pregnancy. *Cancer Prevention Research (Philadelphia, Pa.), 3*, 301–311.
72. Rotunno, M., Sun, X., Figueroa, J., Sherman, M. E., Garcia-Closas, M., Meltzer, P., et al. (2014). Parity-related molecular signatures and breast cancer subtypes by estrogen receptor status. *Breast Cancer Research, 16*, R74.
73. Clarkson, R. W., Wayland, M. T., Lee, J., Freeman, T., & Watson, C. J. (2004). Gene expression profiling of mammary gland development reveals putative roles for death receptors and immune mediators in post-lactational regression. *Breast Cancer Research, 6*, R92–R109.
74. Stein, T., Morris, J. S., Davies, C. R., Weber-Hall, S. J., Duffy, M. A., Heath, V. J., et al. (2004). Involution of the mouse mammary gland is associated with an immune cascade and an acute-phase response, involving LBP, CD14 and STAT3. *Breast Cancer Research, 6*, R75–R91.
75. Liu, Q., Wuu, J., Lambe, M., Hsieh, S. F., Ekbom, A., & Hsieh, C. C. (2002). Transient increase in breast cancer risk after giving birth: Postpartum period with the highest risk (Sweden). *Cancer Causes & Control, 13*, 299–305.
76. Schumacher, A., Heinze, K., Witte, J., Poloski, E., Linzke, N., Woidacki, K., et al. (2013). Human chorionic gonadotropin as a central regulator of pregnancy immune tolerance. *Journal of Immunology, 190*, 2650–2658.
77. Gadi, V. K. (2010). Fetal microchimerism in breast from women with and without breast cancer. *Breast Cancer Research and Treatment, 121*, 241–244.
78. Gadi, V. K. (2009). Fetal microchimerism and cancer. *Cancer Letters, 276*, 8–13.
79. Gadi, V. K., & Nelson, J. L. (2007). Fetal microchimerism in women with breast cancer. *Cancer Research, 67*, 9035–9038.
80. Eun, J. K., Guthrie, K. A., Zirpoli, G., & Gadi, V. K. (2013). In situ breast cancer and microchimerism. *Scientific Reports, 3*, 2192.
81. Boyon, C., Collinet, P., Boulanger, L., Rubod, C., Lucot, J. P., & Vinatier, D. (2011). Fetal microchimerism: Benevolence or malevolence for the mother? *European Journal of Obstetrics, Gynecology, and Reproductive Biology, 158*, 148–152.
82. Kallenbach, L. R., Johnson, K. L., & Bianchi, D. W. (2011). Fetal cell microchimerism and cancer: A nexus of reproduction, immunology, and tumor biology. *Cancer Research, 71*, 8–12.
83. Finak, G., Bertos, N., Pepin, F., Sadekova, S., Souleimanova, M., Zhao, H., et al. (2008). Stromal gene expression predicts clinical outcome in breast cancer. *Nature Medicine, 14*, 518–527.
84. Winslow, S., Leandersson, K., Edsjo, A., & Larsson, C. (2015). Prognostic stromal gene signatures in breast cancer. *Breast Cancer Research, 17*, 23.
85. Nagalla, S., Chou, J. W., Willingham, M. C., Ruiz, J., Vaughn, J. P., Dubey, P., et al. (2013). Interactions between immunity, proliferation and molecular subtype in breast cancer prognosis. *Genome Biology, 14*, R34; Farmer, P., Bonnefoi, H., Anderle, P., Cameron, D., Wirapati, P., Becette, V., et al. (2009). A stroma-related gene signature predicts resistance to neoadjuvant chemotherapy in breast cancer. *Nature Medicine, 15*, 68–74.

86. Mahmoud, S. M., Paish, E. C., Powe, D. G., Macmillan, R. D., Grainge, M. J., Lee, A. H., et al. (2011). Tumor-infiltrating CD8+ lymphocytes predict clinical outcome in breast cancer. *Journal of Clinical Oncology, 29*, 1949–1955.
87. Huang, H., Hara, A., Homma, T., Yonekawa, Y., & Ohgaki, H. (2005). Altered expression of immune defense genes in pilocytic astrocytomas. *Journal of Neuropathology and Experimental Neurology, 64*, 891–901.
88. Lin, W., & Dent, S. Y. (2006). Functions of histone-modifying enzymes in development. *Current Opinion in Genetics & Development, 16*, 137–142.
89. Zuo, T., Tycko, B., Liu, T. M., Lin, H. J., & Huang, T. H. (2009). Methods in DNA methylation profiling. *Epigenomics, 1*, 331–345.
90. Robinson, J. T., Thorvaldsdottir, H., Winckler, W., Guttman, M., Lander, E. S., Getz, G., & Mesirov, J. P. (2011). Integrative genomics viewer. *Nature Biotechnology, 29*, 24–26.
91. Thorvaldsdottir, H., Robinson, J. T., & Mesirov, J. P. (2013). Integrative Genomics Viewer (IGV): High-performance genomics data visualization and exploration. *Briefings in Bioinformatics, 14*, 178–192.
92. Rakyan, V. K., Down, T. A., Thorne, N. P., Flicek, P., Kulesha, E., Graf, S., Tomazou, E. M., Backdahl, L., Johnson, N., Herberth, M., et al. (2008). An integrated resource for genome-wide identification and analysis of human tissue-specific differentially methylated regions (tDMRs). *Genome Research, 18*, 1518–1529.
93. Habas, R., & Dawid, I. B. (2005). Dishevelled and Wnt signaling: Is the nucleus the final frontier? *Journal of Biology, 4*, 2.
94. Aiyar, S. E., Cho, H., Lee, J., & Li, R. (2007). Concerted transcriptional regulation by BRCA1 and COBRA1 in breast cancer cells. *International Journal of Biological Sciences, 3*, 486–492.
95. Wang, X., Goode, E. L., Fredericksen, Z. S., Vierkant, R. A., Pankratz, V. S., Liu-Mares, W., Rider, D. N., Vachon, C. M., Cerhan, J. R., Olson, J. E., et al. (2008). Association of genetic variation in genes implicated in the beta-catenin destruction complex with risk of breast cancer. *Cancer Epidemiology, Biomarkers & Prevention, 17*, 2101–2108.
96. Ratcliffe, M. J., Itoh, K., & Sokol, S. Y. (2000). A positive role for the PP2A catalytic subunit in Wnt signal transduction. *The Journal of Biological Chemistry, 275*, 35680–35683.
97. Cheyette, B. N., Waxman, J. S., Miller, J. R., Takemaru, K., Sheldahl, L. C., Khlebtsova, N., Fox, E. P., Earnest, T., & Moon, R. T. (2002). Dapper, a Dishevelled-associated antagonist of beta-catenin and JNK signaling, is required for notochord formation. *Developmental Cell, 2*, 449–461.
98. Zhao, L., Hart, S., Cheng, J., Melenhorst, J. J., Bierie, B., Ernst, M., Stewart, C., Schaper, F., Heinrich, P. C., Ullrich, A., et al. (2004). Mammary gland remodeling depends on gp130 signaling through Stat3 and MAPK. *The Journal of Biological Chemistry, 279*, 44093–44100.
99. Chang, P. H., Hwang-Verslues, W. W., Chang, Y. C., Chen, C. C., Hsiao, M., Jeng, Y. M., Chang, K. J., Lee, E. Y., Shew, J. Y., & Lee, W. H. (2012). Activation of Robo1 signaling of breast cancer cells by Slit2 from stromal fibroblast restrains tumorigenesis via blocking PI3K/Akt/beta-catenin pathway. *Cancer Research, 72*, 4652–4661.
100. Bertucci, M. C., & Mitchell, C. A. (2013). Phosphoinositide 3-kinase and INPP4B in human breast cancer. *Annals of the New York Academy of Sciences, 1280*, 1–5.
101. Nusse, R., & Varmus, H. (2012). Three decades of Wnts: A personal perspective on how a scientific field developed. *The EMBO Journal, 31*, 2670–2684.
102. Turashvili, G., Bouchal, J., Burkadze, G., & Kolar, Z. (2006). Wnt signaling pathway in mammary gland development and carcinogenesis. *Pathobiology, 73*, 213–223.
103. Toniolo, P., Grankvist, K., Wulff, M., et al. (2010). Human chorionic gonadotropin in pregnancy and maternal risk of breast cancer. *Cancer Research, 70*, 6779–6786.
104. Bernstein, L., Hanisch, R., Sullivan-Halley, J., et al. (1995). Treatment with human chorionic gonadotropin and risk of breast cancer. *Cancer Epidemiology, Biomarkers & Prevention: a publication of the American Association for Cancer Research, cosponsored by the American Society of Preventive Oncology., 4*, 437–440.

105. Tay, L. K., & Russo, J. (1981). 7, 12-Dimethylbenz (a) anthracene (DMBA) induced DNA binding and repair synthesis in susceptible and non-susceptible mammary epithelial cells in culture. *Journal of the National Cancer Institute, 67*, 155–161.
106. Srivastava, P., Russo, J., & Russo, I. H. (1997). Chorionic gonadotropin inhibits rat mammary carcinogenesis through activation of programmed cell death. *Carcinogenesis, 18*, 1799–1808.
107. Medina, D., & Kittrell, F. S. (2003). p53 function is required for hormone-mediated protection of mouse mammary tumorigenesis. *Cancer Research, 63*, 6140–6143.
108. Qi, J., Singh, S., Hua, W.-K., et al. (2015). HDAC8 inhibition specifically targets Inv (16) acute myeloid leukemic stem cells by restoring p53 acetylation. *Cell Stem Cell, 17*, 597–610.
109. Qi, J., Cai, Q., Singh, S., et al. (2014). Inhibition of HDAC8 reactivates p53 and abrogates leukemia stem cell activity in CBFβ-SMMHC associated acute myeloid leukemia. *Blood, 124*, 363–363.
110. Ryu, H. W., Shin, D. H., Lee, D. H., et al. (2017). HDAC6 deacetylates p53 at lysines 381/382 and differentially coordinates p53-induced apoptosis. *Cancer Letters, 391*, 162–171.
111. Thakur, A., Rahman, K. W., Wu, J., et al. (2007). Aberrant expression of X-linked genes RbAp46, Rsk4, and Cldn2 in breast cancer. *Molecular Cancer Research, 5*, 171–181.
112. Sirchia, S. M., Ramoscelli, L., Grati, F. R., et al. (2005). Loss of the inactive X chromosome and replication of the active X in BRCA1-defective and wild-type breast cancer cells. *Cancer Research, 65*, 2139–2146.
113. Wang, X., Lu, G., Li, L., et al. (2014). HUWE1 interacts with BRCA1 and promotes its degradation in the ubiquitin-proteasome pathway. *Biochemical and Biophysical Research Communications, 444*, 290–295.
114. Chen, G. C., Guan, L. S., Yu, J. H., et al. (2001). Rb-associated protein 46 (RbAp46) inhibits transcriptional transactivation mediated by BRCA1. *Biochemical and Biophysical Research Communications, 284*, 507–514.
115. Chen, X., Arciero, C. A., Wang, C., et al. (2006). BRCC36 is essential for ionizing radiation-induced BRCA1 phosphorylation and nuclear foci formation. *Cancer Research, 66*, 5039–5046.
116. Jenuwein, T., & Allis, C. D. (2001). Translating the histone code. *Science, 293*, 1074–1080.
117. Cao, R., Wang, L., Wang, H., et al. (2002). Role of histone H3 lysine 27 methylation in Polycomb-group silencing. *Science, 298*, 1039–1043.
118. Holm, K., Grabau, D., Lovgren, K., et al. (2012). Global H3K27 trimethylation and EZH2 abundance in breast tumor subtypes. *Molecular Oncology, 6*, 494–506.
119. Yan, N., Xu, L., Wu, X., Zhang, L., Fei, X., Cao, Y., & Zhang, F. (2017). GSKJ4, an H3K27me3 demethylase inhibitor, effectively suppresses the breast cancer stem cells. *Experimental Cell Research, 359*, 405–414.

Chapter 9
The Present and Future of Screening in Breast Cancer Prevention

Abstract Mammographic breast density is a strong risk factor for breast cancer, and therefore cancer screening programs are widely implemented to reduce cancer incidence and mortality. Use of hormone therapy (HT) during menopause and mammographic breast density are both associated with an increased risk of breast cancer. An increase in mammographic breast density might result from an absolute increase in fibrous stroma, ductal tissue, or terminal ductal lobular units or from a decrease in fat content with only relative increases in the other elements. Such changes would reflect alterations in proliferation, apoptosis, or a differentiated function. The increase in breast density is associated with the accumulation of target cells for neoplastic transformation providing a biological mechanism why the increase in breast density is a risk factor. Standardization of the mammographic screening has contributed significantly to assess the importance of mammographic density as a predictor of risk. The American College of Radiology (ACR) has established a uniform way for radiologists to describe mammogram findings. The system, called Breast Imaging Reporting and Data system (BI-RADS), includes seven standardized categories, or levels. Digital mammography has been able to make more reproducible and comparable the breast density and be more quantitative in nature. Although there are many software applications, the one that has been shown to be highly accepted in many radiology centers is the VOLPARA (VolparaDataManager®, USA).

Keywords Mammographic breast density · Hormone therapy (HT) · Menopause. Fibrous stroma · Ductal tissue · Terminal ductal lobular units. Proliferation · Apoptosis · Differentiated function · BI-RADSm · VOLPARA

9.1 Mammographic Screening

Mammographic breast density is a strong risk factor for breast cancer [1–3], and therefore cancer screening programs are widely implemented to reduce cancer incidence and mortality. Precise quantitation of breast density has allowed the conclusion from at least 14 studies that mammographic breast density provides a powerful means to predict the risk of breast cancer for a 10- to 15-year period after the

J. Russo, *The Future of Prevention and Treatment of Breast Cancer*,
https://doi.org/10.1007/978-3-030-72815-1_9

mammogram is taken [4–6]. Women older than 55 years whose breast density exceeds 75% have a greater than sixfold increased risk of breast cancer [1, 7]. However, whether breast density is a general marker of susceptibility or is specific to the location of the eventual cancer is unknown [3, 8]. Byrne et al. [7] suggested that these effects are noted for both premenopausal and postmenopausal women of all ages. Findings of studies suggest that higher density is associated with more aggressive and also with in situ tumors [9, 10]. In addition it has been reported that an association between density and risk may be the result of environmental factors [11].

9.2 Hormones and Breast Density

Use of hormone therapy (HT) during menopause and mammographic breast density are both associated with an increased risk of breast cancer [12–15]. Specifically, the prospective, randomized Women's Health Initiative trial found a 26% increase in the relative risk of breast cancer for women using continuous combined estrogen plus progestogen therapy (EPT) over a 5-year period [16]. A large observational study, the Million Women Study, also found a twofold increased risk with EPT [15]. The risk from estrogen alone is not certain based on existing data. The Million Women Study reported a 30% increased relative risk of breast cancer with use of estrogen alone, whereas the prospective Women's Health Initiative trial did not confirm this finding [17–19]. The Nurses' Health Study has reported an increased risk of breast cancer, particularly estrogen receptor- and progesterone receptor-positive invasive carcinomas, with use of estrogen alone for more than 20 years [10]. Women using therapies that reduce breast cancer risk, such as tamoxifen [20, 21] and raloxifene [22–24], experience a decrease in mammographic breast density.

9.3 Biological Basis of Breast Density

An increase in mammographic breast density might result from an absolute increase in fibrous stroma, ductal tissue, or terminal ductal lobular units or from a decrease in fat content with only relative increases in the other elements. Such changes would reflect alterations in proliferation, apoptosis, or differentiated function. The biochemical regulation of breast tissue proliferation and differentiated function in postmenopausal women is poorly understood at present. Inferences from studies in isolated breast cancer cells or co-cultures of fibroblasts and cancer cells allow a focus on several concepts. For example, an increase in fibrous stroma could directly result in increased aromatase, an enzyme known by immunohistochemical analysis and real-time polymerase chain reaction to be present in fibroblasts [25–27]. The resulting increase in production of estradiol could then stimulate epithelial cells to proliferate. Alternatively, the stromal cells produce growth factors such as insulin-like growth factor 1 or proteoglycans, which could act in a paracrine fashion to stimulate epithelial cell proliferation directly [28]. Increased growth factor production, stimulated by the effects of exogenous estrogen, could directly enhance the effects of estradiol.

Harvey et al. [29] have systematically compared mammographic density with histology in women receiving or not receiving HT. To obtain sufficient tissues, they examined the benign areas of mastectomy specimens in women for whom recent mammograms were available. In this study, the authors specifically evaluated mammographic breast density with a computerized method and evaluated correlations with histology as well as estrogen receptor (ER), progesterone receptor (PR), and proliferation indices. There was a statistically significant difference in breast density noted between the HT and non-HT groups ($P < 0.001$) with a median difference in breast density between matched pairs of 23% (54% for HT users and 31% for non-HT users). By using mixed linear effect models, significant clinical predictors of breast density adjusting for HT usage and age included nongravid compared with gravid ($P = 0.002$) and nulliparous compared with parous ($P = 0.034$). Using Spearman partial correlations, increased breast density in women using HT was associated with increased fibrous stroma ($P = 0.020$) and Lob1 ($P = 0.016$) (Table 9.1) but not ducts. Conversely, in nonusers of HT, increased breast density was associated with greater numbers of ducts ($P = 0.022$) but not fibrous stroma or lobules. Overall, increased breast density was significantly associated with greater fibrous stroma ($P = 0.005$) and a greater number of Lob1 ($P = 0.021$), adjusting for HT use and age. With mixed linear effect models, no significant differences in ERs or PRs were observed with increased breast density. Increased Ki67 activities in the ducts ($P = 0.031$) and Lob1 ($P = 0.023$) were also seen in association with increasing breast density for both groups combined [29]

The importance of this report is that increased breast density was significantly associated with Lob1 ($P = 0.016$) as well as Ki67 activity in the ducts ($P = 0.031$) and Lob1 ($P = 0.023$) for both groups. This finding may explain why HT increases the susceptibility of the breast to cancer: it increases the number of target cells or structures to be affected, given that ductal carcinoma in situ starts in the ductules of Lob1 [30, 31]. Dual effects of the hormones are observed here. First, activation of cell proliferation in the breast epithelium of the postmenopausal breast results in

Table 9.1 Partial correlations (Spearman) between increasing mammographic density and histology findings

	HT	No HT	Overall
Fibrous stroma	$r = 0.44$ ($P = 0.020$)	$r = 0.28$ ($P = 0.16$)	$r = 0.38$ (**$P = 0.005$**)
Lobules	$r = 0.46$ (**$P = 0.016$**)	$r = 0.13$ ($P = 0.51$)	$r = 0.31$ (**$P = 0.021$**)
Ducts	$r = 0.04$ ($P = 0.86$)	$r = 0.44$ (**$P = 0.022$**)	$r = 0.20$ ($P = 0.14$)

HT hormone therapy
P values in bold indicate statistical significance

formation of more branching from the ducts, leading to the formation of Lob1, which therefore explains the higher number of Lob1 in the HT group in association with increased breast density. Second, the formation of Lob1 from the ductal structures is associated with the increase in the fibrous stroma because both processes go together [32].

9.4 Mammographic Methods in the Screening for Breast Cancer Risk

As is explained in Sect. 9.2, the increase in breast density is associated with the accumulation of target cells for neoplastic transformation providing a biological mechanism why the increase in breast density is a risk factor. Standardization of the mammographic screening has contributed significantly to assess the importance of mammographic density as a predictor of risk. The American College of Radiology (ACR) has established a uniform way for radiologists to describe mammogram findings. The system, called BI-RADS, includes seven standardized categories, or levels. Each BI-RADS category has a follow-up plan with it to help radiologists and other physicians appropriately manage patient's care. BI-RADS is a quality assurance tool designed to standardize mammography reporting, reduce confusion in breast imaging interpretation, and facilitate outcome monitoring (Table 9.2).

Table 9.2 Radiology assessment—BI-RADS

Final assessment categories			
Category		Management	Likelihood of cancer
0	Need additional imaging or prior examinations	Recall for additional imaging and/or await prior examinations	n/a
1	Negative	Routine screening	Essentially 0%
2	Benign	Routine screening	Essentially 0%
3	Probably benign	Short-interval follow-up (6 month) or continued	>0% but ≤2%
4	Suspicious	Tissue diagnosis	4a. Low suspicion for malignancy (>2% to ≤10%)
			4b. Moderate suspicion for malignancy (>10% to ≤50%)
			4c. High suspicion for malignancy (>50% to <95%)
5	Highly suggestive of malignancy	Tissue diagnosis	≥95%
6	Known biopsy-proven	Surgical excision when clinically appropriate	n/a

Source: The Radiology Assistant, BI-RADS 2013, Harmien Zonderland and Robin Smithuis, the Netherlands

Table 9.3 Comparative methods for determination of breast density

	Qualitative		Quantitative
	Wolfe	BI-RADS	Visual estimation
Description	Visual classification of the mammographic image into four categories based on extent and distribution of the parenchyma, including ducts, nodular, homogeneous densities, and fat	Standardized reporting of visual assessment of mammographic findings by the American College of Radiology BI-RADS. Both breasts are used for the BI-RADS	Radiologist or expert reader subjectively assigns a percentage density corresponding to the proportion of breast that is dense
Categorization	N1 – Completely fatty breast	Category 1: Almost entirely fatty (<25% dense)	<10% 10–<25% 25–<50% 50–<75% >75%
	P1 – Mainly fatty breast with prominent ducts, up to 25% density	Category 2: Scattered fibroglandular densities (25–50% dense)	
	P2 – Prominent ducts, more than 25% density	Category 3: Heterogeneously dense (51–75%)	
	DY – No visible duct, diffuse and extensive nodular density	Category 4: Extremely dense (>75% dense)	
		Category 5: Suggestive of malignancy (>95% dense)	

Adapted from Ref. [8]

In the USA, the Breast Imaging Reporting and Data System (BI-RADS) was developed to standardize mammographic reporting terminology and the assessment and recommendation categories (Table 9.3). BI-RADS density classification is included in this report to inform referring physicians of the decline in sensitivity of mammography with increasing breast density. The five categories of the BI-RADS system are as follows: 1 = almost entirely fatty (<25% dense); 2 = scattered fibro glandular densities (25–50% dense); 3 = heterogeneously dense (51–75% dense); 4 = extremely dense (>75% dense); and 5 = highly suggestive of malignancy (>95% dense). This classification is based on a quantitative assessment, although the categories are not defined by the percentage of density. The advantage of the BI-RADS classification is that it is widely used all over the world and allows analysis of large study populations [33]. In Table 9.3 are depicted the different methods used in the clinical practice.

Digital mammography has been able to make more reproducible and comparable the breast density and be more quantitative in nature. Although there are many software applications, the one that has been shown to be highly accepted in many radiology centers is the Volpara® Data Manager. Basically the principle is that when a mammogram is performed, the x-ray source generates x-rays which penetrate the breast. Those x-rays are attenuated (absorbed) differently by the different types of breast tissue, with fibroglandular tissue attenuating roughly twice the amount of x-rays compared to fat tissue. It is this difference in x-ray attenuation that generates a contrast in each pixel of the image, with fat tissue appearing darker and denser tissue appearing brighter on a mammogram. Volpara Algorithm uses a model of the physics of digital mammography to work backward from the pixel value in the image to the x-ray attenuation between the image detector and the x-ray source. From there, it calculates the types of tissue that must have been present between the detector and the x-ray source. This is a complex process that includes, for example, compensation for beam hardening and scattered radiation. This software provides a sophisticated algorithm that objectively and consistently provides the radiologist with breast composition information which includes BI-RADS-like values as well as area breast density assessment. The software allows to evaluate the pixel dimensions of the detector which are reported in the DICOM header information; the *thickness of fibroglandular and fat tissue at each pixel* can easily be converted into volumes at each pixel (i.e., Tissue Volume = Tissue Thickness × Pixel Width × Pixel Length) [34]. This software analyzes each image within a study and provides values of each screening view, each breast as well as per patient providing: 1-Volume of the breast in cubic centimeters (cc); 2-Volume of fibroglandular tissue in cubic centimeters; 3-Volumetric percentage of fibroglandular tissue (volume of fibroglandular tissue divided by volume of the breast); 4-BI-RADS-like integer score and fractional score; 5-Area of breast density (Abd) in percentage of total breast area, to facilitate study comparison; and 6-Comparison of results to a reference population.

9.5 Evaluation of the Mammographic Screening

Every year, ~46 million individuals are offered cancer screening worldwide, and among them is mammography for breast cancer. It has also become evident that some screening tests, such as mammography, lead to increased cancer incidence rates as a result of over diagnosis—that is, the detection of cancers and cancer precursors that would not have progressed to symptoms or death in the absence of screening. Because over diagnosed lesions cannot be distinguished from lesions that will progress, all patients are treated. Therefore, over diagnosed patients only experience harms and do not gain any benefit from screening according to Kalager and Bretthauser [35].

9.6 Future Tests for Screening

Surgery is curative if the cancer is localized [36]; however once distant metastasis has occurred, surgical excision is rarely curative. In general, it takes several years even 2 to 3 decades for incipient neoplastic lesions to progress to late-stage disease [37–39]. When metastasis has initiated but is not yet evident radiologically, the use of systemic therapy can reduce 50% the course of the disease [40–43].

Therefore, there is a window of opportunity to detect an early event by using markers of risk to develop breast cancer. One of the discussed above has been mammography density, but there are other possibilities like to detect markers in blood like the one used for prostate cancer [44].

However, the blood tests must have very high specificity; otherwise, too many healthy individuals will receive positive test results. Liquid biopsies that detect somatic mutations offer the promise of high specificity because they are based on driver gene mutations that are expected to be found only in abnormal clonal proliferations of cells, such as cancers [45–52]. One important issue against this technology is to determine the tissue of origin in due that the driver mutation must be found in more that on primary site [53–55]. In a publication by Cohen DL et al. in Science [56], they describe a new blood test, called CancerSEEK, that is the use of combined assays for genetic alterations and protein biomarkers and has the capacity not only to identify the presence of relatively early cancers but also to localize the organ of origin of these cancers. This technology offers promising results; however a critical testing and validation process will be required for it to be used efficiently in the clinic.

References

1. Boyd, N. F., Byng, J. W., Jong, R. A., Fishell, E. K., Little, L. E., Miller, A. B., Lockwood, G. A., Tritchler, D. L., & Yaffe, M. J. (1995). Quantitative classification of mammographic densities and breast cancer risk: Results from the Canadian National Breast Screening Study. *Journal of the National Cancer Institute, 87*, 670–675.
2. Boyd, N. F., Martin, L., Jaffe, M. J., & Minkin, S. (2011). Mammographic density and breast cancer risk: Current understanding and future prospects. *Breast Cancer Research, 13*, 223.
3. Masala, G., Ambrogetti, D., Assedi, M., Bendinelli, B., Caini, S., & Palli, D. (2017). Mammographic breast density and breast cancer risk in a Mediterranean population: A nested case-control study in the EPIC Florence cohort. *Breast Cancer Research and Treatment, 164*(2), 467–473. https://doi.org/10.1007/s10549-017-4274-9
4. Harvey, J. A., & Bovbjerg, V. E. (2004). Quantitative assessment of mammographic breast density: Relationship with breast cancer risk. *Radiology, 230*, 29–41.
5. Vacek, P. M., & Geller, B. M. (2004). A prospective study of breast cancer risk using routine mammographic breast density measurements. *Cancer Epidemiology, Biomarkers & Prevention, 13*, 715–722.
6. Nagata, C., Matsubara, T., Fujita, H., et al. (2005). Mammographic density and the risk of breast cancer in Japanese women. *British Journal of Cancer, 92*, 2102–2106.
7. Byrne, C., Schairer, C., Wolfe, J., et al. (1995). Mammographic features and breast cancer risk: Effects with time, age, and menopause status. *Journal of the National Cancer Institute, 87*, 1622–1629.
8. Vachon, C. M., van Gils, C. H., Sellers, T. A., Ghosh, K., Pruthi, S., Brandt, K. R., & Pankratz, V. S. (2007). Mammographic density, breast cancer risk and risk prediction. *Breast Cancer Research, 9*, 217.
9. Kerlikowske, K., & Phipps, A. I. (2011). Breast density influences tumour subtypes and tumour aggressiveness. *Journal of the National Cancer Institute, 103*, 1143–1145.
10. Yaghjyan, L., Colditz, G. A., Collins, L. C., Schnitt, S. J., Rosner, B., Vachon, C., & Tamini, R. M. (2011). Mammographic breast density and subsequent risk of breast cancer in post-

menopausal women according to tumour characteristics. *Journal of the National Cancer Institute, 103*, 1179–1189.
11. Ursin, G., Lillie, E. O., Lee, E., Cockburn, M., Schork, N. J., et al. (2009). The relative importance of genetics and environment on mammographic density. *Cancer Epidemiology, Biomarkers & Prevention, 18*, 102–112.
12. Schairer, C., Lubin, J., Troist, R., Sturgeon, S., Brinton, L., & Hoover, R. (2000). Menopausal estrogen and estrogen-progestin replacement therapy and breast cancer risk. *Journal of the American Medical Association, 283*, 485–491.
13. Persson, I. R., Weiderpass, E., Bergkvist, L., et al. (1999). Risks of breast and endometrial cancer after estrogen and estrogen-progestin replacement. *Cancer Causes & Control, 10*, 253–260.
14. Ross, R. K., Paganini-Hill, A., Wan, P. C., & Pike, M. C. (2000). Effect of hormone replacement therapy on breast cancer risk: Estrogen versus estrogen plus progestin. *Journal of the National Cancer Institute, 92*, 328–332.
15. Beral, V. (2003). Million Women Study Collaborators Breast cancer and hormone-replacement therapy in the Million Women Study. *Lancet, 362*, 419–427.
16. Rossouw, J. E., Anderson, G. L., Prentice, R. L., et al. (2002). Risks and benefits of estrogen plus progestin in healthy postmenopausal women: Principal results from the Women's Health Initiative randomized controlled trial. *Journal of the American Medical Association, 288*, 321–333.
17. Anderson, G. L., Limacher, M., Assaf, A. R., et al. (2004). Effects of conjugated equine estrogen in postmenopausal women with hysterectomy: The Women's Health Initiative randomized control trial. *Journal of the American Medical Association, 291*, 1701–1712.
18. Stefanick, M. L., Anderson, G. L., Margolis, K. L., et al. (2006). Effects of conjugated equine estrogens on breast cancer and mammography screening in postmenopausal women with hysterectomy. *Journal of the American Medical Association, 295*, 1647–1657.
19. Chen, W. Y., Manson, J. E., Hankinson, S. E., et al. (2006). Unopposed estrogen therapy and the risk of invasive breast cancer. *Archives of Internal Medicine, 166*, 1027–1032.
20. Brisson, J., Brisson, B., Cote, G., Maunsell, E., Berube, S., & Robert, J. (2000). Tamoxifen and mammographic breast densities. *Cancer Epidemiology, Biomarkers & Prevention, 9*, 911–915.
21. Cuzick, J., Warwick, J., Pinney, E., Warren, R. M., & Duffy, S. W. (2004). Tamoxifen and breast density in women at increased risk of breast cancer. *Journal of the National Cancer Institute, 96*, 621–628.
22. Christodoulakos, G. E., Lambrinoudaki, I. V., Vourtsi, A. D., Panoulis, K. P., Kelekis, D. A., & Creatsas, G. C. (2002). Mammographic changes associated with raloxifene and tibolone therapy in postmenopausal women: A prospective study. *Menopause, 9*, 110–116.
23. Freedman, M., San Martin, J., O'Gorman, J., et al. (2001). Digitized mammography: A clinical trial of postmenopausal women randomly assigned to receive raloxifene, estrogen, or placebo. *Journal of the National Cancer Institute, 93*, 51–56.
24. Jackson, V. P., San Martin, J. A., Secrest, R. J., et al. (2003). Comparison of the effect of raloxifene and continuous combined hormone therapy on mammographic breast density and breast tenderness in postmenopausal women. *American Journal of Obstetrics and Gynecology, 188*, 389–394.
25. Purohit, A., Ghilchik, M. W., Leese, M. P., Potter, B. V., & Reed, M. J. (2005). Regulation of aromatase activity by cytokines, PGE2 and 2-methoxyoestrone-3-*O*-sulphamate in fibroblasts derived from normal and malignant breast tissues. *The Journal of Steroid Biochemistry and Molecular Biology, 94*, 167–172.
26. Reed, M. J., & Purohit, A. (2001). Aromatase regulation and breast cancer. *Clinical Endocrinology, 54*, 563–571.
27. Singh, A., Purohit, A., Duncan, L. J., Mokbel, K., Ghilchik, M. W., & Reed, M. J. (1997). Control of aromatase activity in breast tumours: The role of the immune system. *The Journal of Steroid Biochemistry and Molecular Biology, 61*, 185–192.

28. Anderson, E., Clarke, R. B., & Howell, A. (1998). Estrogen responsiveness and control of normal human breast proliferation. *Journal of Mammary Gland Biology and Neoplasia, 3*, 23–35.
29. Harvey, J. A., Santen, R. J., Petroni, G. R., Bovbjerg, V., Smolkin, M. A., Sheriff, F., & Russo, J. (2008). Histology changes in the breast with menopausal hormone therapy use: Correlation with breast density, ER, PgR, and proliferation indices. *Menopause, 15*, 67–73.
30. Russo, J., Gusterson, B. A., Rogers, A. E., Russo, I. H., Wellings, S. R., & van Zwieten, M. J. (1990). Comparative study of human and rat mammary tumorigenesis. *Laboratory Investigation, 62*, 244–278.
31. Russo, J., & Russo, I. H. (2004). *Biological and molecular basis of breast cancer*. Heidelberg: Springer-Verlag.
32. Russo, J., Ao, X., Grill, C., & Russo, I. H. (1999). Pattern of distribution of cells positive for estrogen receptor α and progesterone receptor in relation to proliferating cells in the mammary gland. *Breast Cancer Research and Treatment, 53*, 217–227.
33. Kerlikowske, K., Grady, D., Barclay, J., Frankel, S. D., Ominsky, S. H., Sickles, E. A., & Ernster, V. (1998). Variability and accuracy in mammographic interpretation using the American College of Radiology Breast Imaging Reporting and Data System. *Journal of the National Cancer Institute, 1998*(90), 1801–1809.
34. Volpara® Data Manager User Manual Version 1.0. Volpara Solutions Limited support@volparasolutions.commailto; support@volparasolutions.com.; www.volparasolutions.comhttp://www.volparasolutions.com/; http://www.volparasolutions.com/
35. Kalager, M., & Bretthauer, M. (2020). Improving cancer screening programs. *Science, 367*(6474), 143–144.
36. Siegel, R. L., Miller, K. D., & Jemal, A. (2017). Cancer Statistics, 2017. *CA Cancer J Clin, 67*, 7–30.
37. Vogelstein, B., Papadopoulos, N., Velculescu, V. E., Zhou, S., Diaz, L. A., Jr., & Kinzler, K. W. (2013). Cancer genome landscapes. *Science, 339*, 1546–1558.
38. Jones, S., Chen, W. D., Parmigiani, G., Diehl, F., Beerenwinkel, N., Antal, T., Traulsen, A., Nowak, M. A., Siegel, C., Velculescu, V. E., Kinzler, K. W., Vogelstein, B., Willis, J., & Markowitz, S. D. (2008). Comparative lesion sequencing provides insights into tumor evolution. *Proceedings of the National Academy of Sciences of the United States of America, 105*, 4283–4288.
39. Yachida, S., White, C. M., Naito, Y., Zhong, Y., Brosnan, J. A., Macgregor-Das, A. M., Morgan, R. A., Saunders, T., Laheru, D. A., Herman, J. M., Hruban, R. H., Klein, A. P., Jones, S., Velculescu, V., Wolfgang, C. L., & Iacobuzio-Donahue, C. A. (2012). Clinical significance of the genetic landscape of pancreatic cancer and implications for identification of potential long-term survivors. *Clinical Cancer Research, 18*, 6339–6347.
40. Bozic, I., Reiter, J. G., Allen, B., Antal, T., Chatterjee, K., Shah, P., Moon, Y. S., Yaqubie, A., Kelly, N., Le, D. T., Lipson, E. J., Chapman, P. B., Diaz, L. A., Jr., Vogelstein, B., & Nowak, M. A. (2013). Evolutionary dynamics of cancer in response to targeted combination therapy. *eLife, 2*, e00747.
41. Semrad, T. J., Fahrni, A. R., Gong, I. Y., & Khatri, V. P. (2015). Integrating chemotherapy into the management of oligometastatic colorectal cancer: Evidence-based approach using clinical trial findings. *Annals of Surgical Oncology, 22*(Suppl 3), S855–S862.
42. Moertel, C. G., Fleming, T. R., Macdonald, J. S., Haller, D. G., Laurie, J. A., Tangen, C. M., Ungerleider, J. S., Emerson, W. A., Tormey, D. C., Glick, J. H., Veeder, M. H., & Mailliard, J. A. (1995). Fluorouracil plus levamisole as effective adjuvant therapy after resection of stage III colon carcinoma: A final report. *Annals of Internal Medicine, 122*, 321–326.
43. Huang, A. C., Postow, M. A., Orlowski, R. J., Mick, R., Bengsch, B., Manne, S., Xu, W., Harmon, S., Giles, J. R., Wenz, B., Adamow, M., Kuk, D., Panageas, K. S., Carrera, C., Wong, P., Quagliarello, F., Wubbenhorst, B., D'Andrea, K., Pauken, K. E., Herati, R. S., Staupe, R. P., Schenkel, J. M., McGettigan, S., Kothari, S., George, S. M., Vonderheide, R. H., Amaravadi, R. K., Karakousis, G. C., Schuchter, L. M., Xu, X., Nathanson, K. L., Wolchok, J. D.,

Gangadhar, T. C., & Wherry, E. J. (2017). T-cell invigoration to tumor burden ratio associated with anti-PD-1 response. *Nature, 545*, 60–65.

44. Pinsky, P. F., Prorok, P. C., & Kramer, B. S. (2017). Prostate cancer screening – a perspective on the current state of the evidence. *The New England Journal of Medicine, 376*, 1285–1289.
45. Bettegowda, C., Sausen, M., Leary, R. J., Kinde, I., Wang, Y., Agrawal, N., Bartlett, B. R., Wang, H., Luber, B., Alani, R. M., Antonarakis, E. S., Azad, N. S., Bardelli, A., Brem, H., Cameron, J. L., Lee, C. C., Fecher, L. A., Gallia, G. L., Gibbs, P., Le, D., Giuntoli, R. L., Goggins, M., Hogarty, M. D., Holdhoff, M., Hong, S. M., Jiao, Y., Juhl, H. H., Kim, J. J., Siravegna, G., Laheru, D. A., Lauricella, C., Lim, M., Lipson, E. J., Marie, S. K., Netto, G. J., Oliner, K. S., Olivi, A., Olsson, L., Riggins, G. J., Sartore-Bianchi, A., Schmidt, K., Shih, I. M., Oba-Shinjo, S. M., Siena, S., Theodorescu, D., Tie, J., Harkins, T. T., Veronese, S., Wang, T. L., Weingart, J. D., Wolfgang, C. L., Wood, L. D., Xing, D., Hruban, R. H., Wu, J., Allen, P. J., Schmidt, C. M., Choti, M. A., Velculescu, V. E., Kinzler, K. W., Vogelstein, B., Papadopoulos, N., & Diaz, L. A., Jr. (2014). Detection of circulating tumor DNA in early- and late-stage human malignancies. *Sci Transl Med, 6*, 224ra24.
46. Haber, D. A., & Velculescu, V. E. (2014). Blood-based analyses of cancer: Circulating tumor cells and circulating tumor DNA. *Cancer Discovery, 4*, 650–661.
47. Dawson, S. J., Tsui, D. W., Murtaza, M., Biggs, H., Rueda, O. M., Chin, S. F., Dunning, M. J., Gale, D., Forshew, T., Mahler-Araujo, B., Rajan, S., Humphray, S., Becq, J., Halsall, D., Wallis, M., Bentley, D., Caldas, C., & Rosenfeld, N. (2013). Analysis of circulating tumor DNA to monitor metastatic breast cancer. *The New England Journal of Medicine, 368*, 1199–1209.
48. Wang, Y., Springer, S., Mulvey, C. L., Silliman, N., Schaefer, J., Sausen, M., James, N., Rettig, E. M., Guo, T., Pickering, C. R., Bishop, J. A., Chung, C. H., Califano, J. A., Eisele, D. W., Fakhry, C., Gourin, C. G., Ha, P. K., Kang, H., Kiess, A., Koch, W. M., Myers, J. N., Quon, H., Richmon, J. D., Sidransky, D., Tufano, R. P., Westra, W. H., Bettegowda, C., Diaz, L. A., Jr., Papadopoulos, N., Kinzler, K. W., Vogelstein, B., & Agrawal, N. (2015). Detection of somatic mutations and HPV in the saliva and plasma of patients with head and neck squamous cell carcinomas. *Sci Transl Med, 7*, 293ra104.
49. Forshew, T., Murtaza, M., Parkinson, C., Gale, D., Tsui, D. W., Kaper, F., Dawson, S. J., Piskorz, A. M., Jimenez-Linan, M., Bentley, D., Hadfield, J., May, A. P., Caldas, C., Brenton, J. D., & Rosenfeld, N. (2012). Noninvasive identification and monitoring of cancer mutations by targeted deep sequencing of plasma DNA. *Sci Transl, 4*, 136ra68.
50. Abbosh, C., Birkbak, N. J., Wilson, G. A., Jamal-Hanjani, M., Constantin, T., Salari, R., Le Quesne, J., Moore, D. A., Veeriah, S., Rosenthal, R., Marafioti, T., Kirkizlar, E., TBK, W., McGranahan, N., Ward, S., Martinson, L., Riley, J., Fraioli, F., Al Bakir, M., Grönroos, E., Zambrana, F., Endozo, R., Bi, W. L., Fennessy, F. M., Sponer, N., Johnson, D., Laycock, J., Shafi, S., Czyzewska-Khan, J., Rowan, A., Chambers, T., Matthews, N., Turajlic, S., Hiley, C., Lee, S. M., Forster, M. D., Ahmad, T., Falzon, M., Borg, E., Lawrence, D., Hayward, M., Kolvekar, S., Panagiotopoulos, N., Janes, S. M., Thakrar, R., Ahmed, A., Blackhall, F., Summers, Y., Hafez, D., Naik, A., Ganguly, A., Kareht, S., Shah, R., Joseph, L., Marie Quinn, A., Crosbie, P. A., Naidu, B., Middleton, G., Langman, G., Trotter, S., Nicolson, M., Remmen, H., Kerr, K., Chetty, M., Gomersall, L., Fennell, D. A., Nakas, A., Rathinam, S., Anand, G., Khan, S., Russell, P., Ezhil, V., Ismail, B., Irvin-Sellers, M., Prakash, V., Lester, J. F., Kornaszewska, M., Attanoos, R., Adams, H., Davies, H., Oukrif, D., Akarca, A. U., Hartley, J. A., Lowe, H. L., Lock, S., Iles, N., Bell, H., Ngai, Y., Elgar, G., Szallasi, Z., Schwarz, R. F., Herrero, J., Stewart, A., Quezada, S. A., Peggs, K. S., Van Loo, P., Dive, C., Lin, C. J., Rabinowitz, M., HJWL, A., Hackshaw, A., Shaw, J. A., Zimmermann, B. G., TRACERx consortium, PEACE consortium, & Swanton, C. (2017). Phylogenetic ctDNA analysis depicts early-stage lung cancer evolution. *Nature, 545*, 446–451.
51. Beddowes, E., Sammut, S. J., Gao, M., & Caldas, C. (2017). Predicting treatment resistance and relapse through circulating DNA. *Breast, 34*(Suppl 1), S31–S35.
52. Phallen, J., Sausen, M., Adleff, V., Leal, A., Hruban, C., White, J., Anagnostou, V., Fiksel, J., Cristiano, S., Papp, E., Speir, S., Reinert, T., Orntoft, M. W., Woodward, B. D., Murphy, D.,

Parpart-Li, S., Riley, D., Nesselbush, M., Sengamalay, N., Georgiadis, A., Li, Q. K., Madsen, M. R., Mortensen, F. V., Huiskens, J., Punt, C., van Grieken, N., Fijneman, R., Meijer, G., Husain, H., Scharpf, R. B., Diaz, L. A., Jr., Jones, S., Angiuoli, S., Ørntoft, T., Nielsen, H. J., Andersen, C. L., & Velculescu, V. E. (2017). Direct detection of early-stage cancers using circulating tumor DNA. *Sci Transl Med, 9*, eaan 2415.

53. Cree, I. A., Uttley, L., Buckley Woods, H., Kikuchi, H., Reiman, A., Harnan, S., Whiteman, B. L., Philips, S. T., Messenger, M., Cox, A., Teare, D., Sheils, O., & Shaw, J. (2017). UK Early Cancer Detection Consortium. The evidence base for circulating tumour DNA blood-based biomarkers for the early detection of cancer: a systematic mapping review. *BMC Cancer, 17*, 697.
54. Cohen, J. D., Javed, A. A., Thoburn, C., Wong, F., Tie, J., Gibbs, P., Schmidt, C. M., Yip-Schneider, M. T., Allen, P. J., Schattner, M., Brand, R. E., Singhi, A. D., Petersen, G. M., Hong, S. M., Kim, S. C., Falconi, M., Doglioni, C., Weiss, M. J., Ahuja, N., He, J., Makary, M. A., Maitra, A., Hanash, S. M., Dal Molin, M., Wang, Y., Li, L., Ptak, J., Dobbyn, L., Schaefer, J., Silliman, N., Popoli, M., Goggins, M. G., Hruban, R. H., Wolfgang, C. L., Klein, A. P., Tomasetti, C., Papadopoulos, N., Kinzler, K. W., Vogelstein, B., & Lennon, A. M. (2017). Combined circulating tumor DNA and protein biomarker-based liquid biopsy for the earlier detection of pancreatic cancers. *Proc Natl Acad Sci USA, 114*, 10202–10207.
55. Bardelli, A., & Pantel, K. (2017). Liquid Biopsies, What We Do Not Know (Yet). *Cancer Cell, 31*, 172–179.
56. Cohen, J. D., Li, L., Wang, Y., Thoburn, C., Afsari, B., Danilova, L., Douville, C., Javed, A. A., Wong, F., Mattox, A., Hruban, R. H., Wolfgang, C. L., Goggins, M. G., Dal Molin, M., Wang, T. L., Roden, R., Klein, A. P., Ptak, J., Dobbyn, L., Schaefer, J., Silliman, N., Popoli, M., Vogelstein, J. T., Browne, J. D., Schoen, R. E., Brand, R. E., Tie, J., Gibbs, P., Wong, H. L., Mansfield, A. S., Jen, J., Hanash, S. M., Falconi, M., Allen, P. J., Zhou, S., Bettegowda, C., Diaz, L. A., Jr., Tomasetti, C., Kinzler, K. W., Vogelstein, B., Lennon, A. M., & Papadopoulos, N. (2018). Detection and localization of surgically resectable cancers with a multi-analyte blood test. *Science, 359*, 926–930.

Chapter 10
A Vision of the Future

Abstract *Although poverty is an important root for our human problems, lack of education is the original cause of our social problems. Therefore, to educate people is to empower them to solve all the main problems of our human race, and each of us has a role in this aim.* It is known that lifestyle factors play an important role in cancer development. One of the most important lifestyle practices to improve breast cancer survival chances is maintaining a healthful body weight. Being overweight or obese increases the risk of breast cancers in postmenopausal women. Increase in estrogen levels has been associated with risk of breast cancer, and alcohol has been shown to increase estrogen levels. Being physically active lowers the risk of breast cancer recurrence and mortality, decreases treatment-related symptoms, and increases quality of life.

Major medical and governmental organizations recommend mammographic screening for all women starting between the ages of 40 and 50. In the USA and the UK combined, over 42 million exams are performed each year. However, interpretation of these images is challenging. Computer-aided detection (CAD) software for mammography was introduced in the 1990s, and several assistive tools have been approved for medical use. The use of Artificial Intelligence (AI) is in the stage to overcome the shortcoming of the present system of mammographic screening that will allow to improve the performance of human experts on medical-image analysis. Although there is not a blood test for determining cancer risk besides the genetic testing of germ cell mutations, the aim of the future of breast cancer prevention or at least early breast cancer detection is in developing blood test that can detect early markers of disease susceptibility or risk to develop a neoplasia. In chapter 6 we have described a potential test called CancerSEEK that is able to detect markers of cancer development. Liquid biopsies have emerged as a molecular analysis of blood specimens for cancer screening; however cancer gene mutations alone are insufficient to identify the tissue of origin for a given cancer signal in the blood because similar mutations are present in multiple different cancers. Pharmaceutical companies are spending billions of dollars a year for developing few successful drugs, and therefore it is reasonable that the new frontier of AI my help to solve this difficult and intricate problem. The potential use of AI includes not only drug discovery and development but how these drugs are clinically validated and ultimately used in the clinical setup. Another important application of AI is in the evaluation of combination therapy. It has been shown that combination cancer therapies aim to improve the probability and magnitude of therapeutic responses and reduce the likelihood of

J. Russo, *The Future of Prevention and Treatment of Breast Cancer*,
https://doi.org/10.1007/978-3-030-72815-1_10

acquired resistance in an individual patient. Electronic health records (EHRs) are digital versions of the paper charts in which doctors used to record patients' visits, laboratory results, and other important medical information that were supposed to transform the practice of medicine, and AI could be an important tool for improving this area of medicine. Women who carry germ line mutations in either *BRCA1* or *BRCA2* are at a greatly increased risk of breast as well as ovarian cancer. Interventions, such as bilateral mastectomy or breast MRI screening, are available to women who carry a deleterious *BRCA1* or *BRCA2* mutation to lower their chances of being diagnosed with or dying from cancer, but these approaches are far from perfect, especially in younger women. Given the limitations of current strategies for breast cancer prevention, i.e., screening and prophylactic surgery in *BRCA1/2* carriers, it is important to develop new preventative approaches. To regulate the *BRCA* genes encode nuclear proteins that play a role in maintaining genomic stability, promoting DNA repair that leads to tumor suppression will be the optimal solution.

Keywords Body weight · Alcohol consumption · Exercise · Artificial intelligence · Mammographic screening · Drug discovery

10.1 Lay Down of a Vision

"Although poverty is an important root for our human problems, lack of education is the original cause of our social problems. Therefore, to educate people is to empower them to solve all the main problems of our human race, and each of us has a role in this aim." This was the last sentence written in the book *Memoirs of A Cancer Researcher* [1]. The vision of this concept is that economic resources need to be directed to education of our young and teenager population in order that they acquire a strong analytical capability to see the world in a humanistic perspective. To educate the body and mind and give a sense of spiritual value and transcendence to the teaching of their life will provide the basic understanding in human and social interaction and from my medical point of view a better understanding of their body.

The deep understanding of how our body works and interact with the environment will make us more respectful not only of ourselves but also in the interrelation with other human beings and with our surrounding. Among the education that they must receive is how our endocrine system and reproductive apparatus works. In my vision of a world without breast cancer the very simplistic initial condition is addressing the education to women from their youngest age.

In this chapter I will outline some the preventive strategies that we can use in our public health system and are part of a future without cancer. Some of these measures can be immediately applied other will require more time and development for been able to be used in all the sections of our society. Preventive strategies as we see them at the present time are those related to lifestyle factors, screening program, development of more efficient drugs that can be used at early stages of breast cancer preventing the advance of the disease like metastasis that cannot be easily controlled, and finally a better system of health records that can be safely used.

Another preventive measure against breast cancer is to use of a natural hormone, human chorionic gonadotropin(hCG) that is able to change the chromatin structure of the breast cells making those cells refractory to neoplastic transformation [1].

10.2 Lifestyle Factors and Risk

It is known that lifestyle factors play an important role in cancer development. One of the most important lifestyle practices to improve breast cancer survival chances is maintaining a healthful body weight. Being overweight or obese increases the risk of breast cancers in postmenopausal women. Studies show that women who gained 21–30 pounds since age 18 were 40% more likely to develop breast cancer than women who had not gained more than 5 pounds, and women who gained 70 pounds doubled their risk [2–6]. The relationship between obesity and breast cancer in premenopausal women has been shown to be different than those at post-menopause [7].

Although the exact mechanism why obesity increases the risk of breast cancer is not known, there are some evidence that high levels of insulin-like growth factor (IGF-1) in the circulation leads to inflammation which may explain the link between obesity and cancer. Therefore, the more overweight a person is, the greater the level of inflammatory signals takes place in the body [8].

Increase in estrogen levels has been associated with risk of breast cancer, and alcohol has been shown to increase estrogen levels. For ER-positive breast cancer survivors, studies suggest risk of recurrence increases when a woman has more than one or two drinks per week. All women, but especially those who drink, should consume folate-rich foods: leafy greens, legumes, and enriched whole-grain products are good sources [9]. Women who eat a minimum of five servings of vegetables and fruit per day combined with regular physical activity roughly 150 minutes of brisk walking weekly reduce their risk of recurrence even if they have the BRCA mutations [9]. Physical activity over a few months can turn genes on and off. Just 150 minutes of brisk walking a week lowered overall cancer risk. The most notable changes have been seen in inflammation, immune function, DNA repair, and hormonal pathways. Being physically active lowers the risk of breast cancer recurrence and mortality, decreases treatment-related symptoms, and increases quality of life. Activities considered moderate include walking, biking, and even housework and gardening [9].

Besides diet, exercise, and alcohol usage, exposure to light at night and the electromagnetic energy fields generated from cell phones, night lights, and electrical devices by disrupting melatonin production and increasing estrogen levels may also increase the risk to breast cancer.

The role of education starting at early age should emphasize in these basic principles of Mental and Physical Hygiene.

10.3 Mammographic Screening

Major medical and governmental organizations recommend mammographic screening for all women starting between the ages of 40 and 50 [10–12]. In the USA and the UK combined, over 42 million exams are performed each year [13, 14]. However, interpretation of these images is challenging [15–18]. Computer-aided detection (CAD) software for mammography was introduced in the 1990s, and several assistive tools have been approved for medical use [19]. One advance system designated Volpara has been described in Chap. 6.

The use of Artificial Intelligence is in the stage to overcome the shortcoming of the present system of mammographic screening that will allow to improve the performance of human experts on medical-image analysis [20–24]. In a recent reported study [25] it has shown that artificial intelligence (AI) system outperforms radiologists on a clinically relevant task of breast cancer identification. This is a large scale of the data sets used for training and subsequently validating the AI algorithm [26]. Mammograms for 25,856 women in the UK and 3097 women in the USA were used to train the AI system. The system was then used to identify the presence of breast cancer in mammograms of women who were known to have had either biopsy-proven breast cancer or normal follow-up imaging results at least 365 days later. These results held across two large datasets in the UK and the USA that are representative of different screening populations and practices. Of great interest is that the scores of the AI system stratified cases better than the BI-RADS ratings (see Chap. 6) that were assigned by each of the radiologists. A shortcoming of this study [25] is that the vast majority of images used in this study were acquired on devices made by Hologic. Therefore, a broadening spectrum of devices must be used before AI have stronghold in the clinical practice. However, in the long range AI could play an important role in mammographic screening by helping reduce recall rates and unnecessary biopsies. It could also help in detecting cancers earlier than the present system.

In McKinney and colleagues' study [25] the authors suggest that an additional advantage of the use of AI in mammographic screening is that the system could obviate the need for double reading in 88% of UK screening cases, while maintaining a similar level of accuracy to the standard protocol.

A word of caution is to learn with the experience gained when the computer-aided detection (CAD) of breast cancer was introduced that showed great promise in experimental testing, but fell short in real-world settings [16]. The use of AI is in the horizon, but careful control clinical trial needs to be done in order to avoid false positive in which the AI detect a lesion and the radiologist does not. For most women, screening identifies no abnormalities, and when there is still no evidence of cancer 1 year later, this is classified as a true negative. In addition, borderline and complex cases will have required to be incorporated in the validated data sets that can be cross reference with the clinical records of that specific woman.

10.4 Blood Tests

Although there is not a blood test for determining cancer risk besides the genetic testing of germ cell mutations, the aim of the future of breast cancer prevention or at least early breast cancer detection is in developing blood test that can detect early markers of disease susceptibility or risk to develop a neoplasia. In Chap. 6 we have described a potential test called CancerSEEK [27] that is able to detect markers of cancer development.

Liquid biopsies have emerged as a molecular analysis of blood specimens for cancer screening; however cancer gene mutations alone are insufficient to identify the tissue of origin for a given cancer signal in the blood because similar mutations are present in multiple different cancers. In the CancerSEEK [27] plasma-based sequencing of 16 cancer genes generates an Omega score that is combined with eight cancer-associated serum proteins to derive a probability for having any of eight different types of cancer. A machine learning algorithm then integrates these data with 31 additional serum proteins and patient gender to predict the tissue of origin. This test is the beginning of things to come by providing a platform in which direction the development of blood testing should go.

10.5 Targeted Drugs

Pharmaceutical companies are spending billions of dollars a year for developing few successful drugs, and therefore it is reasonable that the new frontier of AI my help to solve this difficult and intricate problem. The potential use of AI includes not only drug discovery and development but how these drugs are clinically validated and ultimately used in the clinical setup [28, 29]. The idea is to create a system that feeding the machine with the molecule with potential to be a drug the software develops its own computational approaches for producing information on how the molecule will work in specific target.

AI platforms are ranging from machine learning to neural networks for accelerating drug discovery, identifying biomarkers that match patients to clinical trials, and personalize the cancer therapy by using patient's own data [30, 31]. The main problem is to identify those drugs that are toxic and focus on candidates that could be possible to be tested in clinical trials and finally be approved by the Food and Drug Administration.

At the present time few AI-discovered drugs are actually in the human-testing pipeline; however, none has begun phase 3 human trials [32]. A very important problem, for example, is to determine if the drug inhibits the production of CYP450 that can render such a drug toxic in humans. P450 is a set of enzymes mostly produced in the liver involved in breaking down chemicals and preventing them from building up to dangerous levels in the bloodstream. In a report of Lin et al. [33] using CRISPR-Cas9 mutagenesis allowed then validation of the mechanism of action of cancer drugs in the preclinical setting facilitating their further use in the clinic.

Another important application of AI is in the evaluation of combination therapy. It has been shown that combination cancer therapies aim to improve the probability and magnitude of therapeutic responses and reduce the likelihood of acquired resistance in an individual patient. As it has been pointed out [34] drug combinations may exhibit additivity or synergy in pre-clinical models, but the patient-to-patient variability and low cross-resistance make independent action the dominant mechanism in clinical populations. Therefore, the use AI can help to clarify this issue more efficiently as it has been demonstrated by Rashid et al. [35] showing that optimizing drug combinations against multiple myeloma can be achieved using a quadratic phenotypic optimization platform (QPOP). Basically this platform uses quadratic relationships represented by parabolas to visually correlate a set of inputs (e.g., drugs and doses) with optimal outputs (e.g., preclinical tumor reduction with minimal toxicity). This correlation markedly reduces the number of experiments and data needed to identify the drugs and doses that optimize combination therapy design.

The AI can provide accurate information and very fast in how the possibilities that a molecule targets a receptor, for example, but it is difficult with the data available how the molecule will pass the intestinal system to the bloodstream without being immediately broken down by the liver; without disrupting other organs; and incapacitating any of the thousands of other proteins in the human body. To solve this problem, it is required more global genetic data about variations in proteins, so that the AI software can specify which patients the candidate drugs would work best on. This crucial step will cut down the number of years required for the drug to get use in the clinic.

An additional compound factor is to identify the molecular target of a disease; for this purpose genomics, proteomics, metabolomics, lipidomics, and microbiomic data are used to identify targets and run through a set of deep-learning programs that search for any differences between non-disease and disease states. The identification of a protein or set of proteins that may be associated to a disease or more importantly that have impact on the disease is the final goal. However, the idea that only one protein is responsible for the disease needs to be carefully considered because the driver of the neoplastic process can be a cascade of molecular events triggering more than one gene or protein.

A silver line in this complexity in the use of AI in the drug discovery that may help in prevention, diagnosis, prognosis, and cure of breast cancer is that it has created the needs of a partnership between large Pharma companies and according to David H. Freedman [25] at least 20 separate partnerships have been reported between the major companies and AI-drug-discovery tech companies.

10.6 Health Records

Electronic health records (EHRs) are digital versions of the paper charts in which doctors used to record patients' visits, laboratory results, and other important medical information were supposed to transform the practice of medicine. The EHRs were implemented in 2009, but it has not been as impacting as it was supposed to be, because of the continuous complaints by physicians on the clunky interfaces and time-consuming data entry. However, this initial effort can be useful for applying the next step, that is, the use of AI. Artificial intelligence in the form of machine learning which allows computers to identify patterns in data and draw conclusions on their own might be able to help overcome the obstacles encountered with EHRs and unlock their potential for making predictions and improving patient care as has been stated by C Wylliard [36].

To accomplish this objective computer scientists start with. Algorithms and train learning machines using real-world examples with known outcomes. However, most of the data found in EHRs are not ready to be fed into an algorithm. A massive amount of curation has to occur first [36].

10.7 Basis of a New Preventive Strategy in Hereditary Breast Cancer

Women who carry germ line mutations in either *BRCA1* or *BRCA2* are at a greatly increased risk of breast as well as ovarian cancer. For mutations in *BRCA1 or BRCA2* the estimated average risk of breast cancer ranges from 57–65% and 35–57%, respectively [37]. Interventions, such as bilateral mastectomy or breast MRI screening, are available to women who carry a deleterious *BRCA1* or *BRCA2* mutation to lower their chances of being diagnosed with or dying from cancer, but these approaches are far from perfect, especially in younger women [38–41].

Given the limitations of current strategies for breast cancer prevention, i.e., screening and prophylactic surgery in *BRCA1/2* carriers [38–42], it is important to develop new preventative approaches.

The *BRCA* genes encode nuclear proteins that play a role in maintaining genomic stability, promoting DNA repair that leads to tumor suppression [43–45]. Decreased activity of *BRCA* genes, whether mutated or epigenetically inhibited, leads to reduced DNA repair and eventually cancer. In this stage of knowledge, we propose an alternative strategy based on studies performed in rodents and humans showing that r-hCG that has the full functional structure of the natural hormone with both alpha and beta chains intact confers the same degree of protection (95%) from mammary cancer development as a full-term pregnancy [46–53]. In our ongoing clinical trial (https://clinicaltrials.gov/beta/with the Clinical trials Gov identifier:NGT00700778) we are aiming to show that the use of recombinant human chorionic gonadotropin (r-hCG at a dose of 250 μg intramuscularly three times a week for the period of 3 months in nulliparous women 19 to 29 years of age and carrier of *BRCA* mutations) will induce the same genomic signature induced in postmenopausal parous women [49–53].

In our laboratory, we will be able to develop novel molecular tools, such as liquid-based biopsy microfluidic platforms and breast EV-associated biomarker assays, to evaluate the effect of r-hCG in a non-invasive assay. In addition, we are aiming to develop a r-hCG depot system of delivery using r-hCG in microspheres containing PEG/PBT. A second line of studies is the prediction and design of small peptides that will mimic the native hormone activation by evaluating which peptides are activating the CG/LH-R. The functional action of both short peptides and a peptide chimera in which both alpha and beta chains are incorporated will be evaluated in the in vitro system followed by in vivo studies using the conditional mutant mouse $Brca1^{F/F}$; $p53^{F/F}$ [54]. This will be followed by establishing safety and pharmacokinetics of the optimal hCG peptide before phase 1 trials in human volunteers.

References

1. Russo, J. (2019). *Memoirs of a cancer researcher*. Singapore: World Publishing.
2. Suzuki, R., Saji, S., & Toi, M. (2012). Impact of body mass index on breast cancer in accordance with the life-stage of women. *Frontiers in Oncology, 2*, 123.
3. World Cancer Research Fund. (2017). *International systematic literature review, the associations between food, nutrition and physical activity and the risk of breast cancer: Continuous update project*. London: World Cancer Research Fund.
4. Premenopausal Breast Cancer Collaborative Group, Schoemaker, M. J., Nichols, H. B., et al. (2018). Association of body mass index and age with subsequent breast cancer risk in premenopausal women. *JAMA Oncol, 4*, e181771.
5. Baer, H. J., Tworoger, S. S., Hankinson, S. E., et al. (2010). Body fatness at young ages and risk of breast cancer throughout life. *American Journal of Epidemiology, 171*, 1183–1194.
6. Colditz, G. A., Bohlke, K., & Berkey, C. S. (2014). Breast cancer risk accumulation starts early: Prevention must also. *Breast Cancer Research and Treatment, 145*, 567–579.
7. Schoemaker, M. J., Nichols, H. B., Wright, L. B., Brook, M. N., Jones, M. E., O'Brien, K. M., Adami, H. O., Baglietto, L., Bernstein, L., Bertrand, K. A., Boutron-Ruault, M. C., Chen, Y., Connor, A. E., Dossus, L., Eliassen, A. H., Giles, G. G., Gram, I. T., Hankinson, S. E., Kaaks, R., Key, T. J., Kirsh, V. A., Kitahara, C. M., Larsson, S. C., Linet, M., Ma, H., Milne, R. L., Ozasa, K., Palmer, J. R., Riboli, E., Rohan, T. E., Sacerdote, C., Sadakane, A., Sund, M., Tamimi, R. M., Trichopoulou, A., Ursin, G., Visvanathan, K., Weiderpass, E., Willett, W. C., Wolk, A., Zeleniuch-Jacquotte, A., Sandler, D. P., & Swerdlow, A. J. (2020). Adult weight change and premenopausal breast cancer risk: A prospective pooled analysis of data from 628,463 women. *Int J Cancer, 147*(5), 1306–1314. https://doi.org/10.1002/ijc.32892
8. Wilding, J. P. H. (2020). Endocrine testing in obesity. *European Journal of Endocrinology, 182*, C13–C15.
9. Akdeniz, D., Klaver, M. M., Smith, C. Z. A., Koppert, L. B., & Hooning, M. J. (2020). The impact of lifestyle and reproductive factors on the risk of a second new primary cancer in the contralateral breast: A systematic review and meta-analysis. *Cancer Causes & Control, 31*(5), 403–416. https://doi.org/10.1007/s10552-020-01284-2
10. Lee, C. H., et al. (2010). Breast cancer screening with imaging: recommendations from the Society of Breast Imaging and the ACR on the use of mammography, breast MRI, breast

ultrasound, and other technologies for the detection of clinically occult breast cancer. *Journal of the American College of Radiology, 7*, 18–27.
11. Oeffinger, K. C., et al. (2015). Breast cancer screening for women at average risk: 2015 guideline update from the American Cancer *Society*. *Journal of the American Medical Association, 314*, 1599–1614.
12. Siu, A. L. (2016). Screening for breast cancer: U.S. Preventive Services Task Force recommendation statement. *Annals of Internal Medicine, 164*, 279–296.
13. Center for Devices & Radiological Health. *MQSA National Statistics* (US Food and Drug Administration, 2019. Accessed 16 July 2019). http://www.fda.gov/radiation-emittingproducts/mqsa-insights/mqsa-national-statistics.
14. Cancer Research UK. *Breast Screening* (CRUK, 2017. Accessed 26 July 2019). https://www.cancerresearchuk.org/about-cancer/breast-cancer/screening/breast-screening.
15. Elmore, J. G., et al. (2009). Variability in interpretive performance at screening mammography and radiologists' characteristics associated with accuracy. *Radiology, 253*, 641–651.
16. Lehman, C. D., et al. (2015). Diagnostic accuracy of digital screening mammography with and without computer-aided detection. *JAMA Internal Medicine, 175*, 1828–1837.
17. Tosteson, A. N. A., et al. (2014). Consequences of false-positive screening mammograms. *JAMA Internal Medicine, 174*, 954–961.
18. Houssami, N., & Hunter, K. (2017). The epidemiology, radiology and biological characteristics of interval breast cancers in population mammography screening. *NPJ Breast Cancer, 3*, 12.
19. Rao, V. M., et al. (2010). How widely is computer-aided detection used in screening and diagnostic mammography? *Journal of the American College of Radiology, 7*, 802–805.
20. Gulshan, V., et al. (2016). Development and validation of a deep learning algorithm for detection of diabetic retinopathy in retinal fundus photographs. *Journal of the American Medical Association, 316*, 2402–2410.
21. Esteva, A., et al. (2017). Dermatologist-level classification of skin cancer with deep neural networks. *Nature, 542*, 115–118.
22. De Fauw, J., et al. (2018). Clinically applicable deep learning for diagnosis and referral in retinal disease. *Nature Medicine, 24*, 1342–1350.
23. Ardila, D., et al. (2019). End-to-end lung cancer screening with three-dimensional deep learning on low-dose chest computed tomography. *Nature Medicine, 25*, 954–961.
24. Topol, E. J. (2019). High-performance medicine: the convergence of human and artificial intelligence. *Nature Medicine, 25*, 44–56.
25. McKinney, S. M., Sieniek, M., Godbole, V., Godwin, J., Antropova, N., Ashrafian, H., Back, T., Chesus, M., Corrado, G. C., Darzi, A., Etemadi, M., Garcia-Vicente, F., Gilbert, F. J., Halling-Brown, M., Hassabis, D., Jansen, S., Karthikesalingam, A., Kelly, C. J., King, D., Ledsam, J. R., Melnick, D., Mostofi, H., Peng, L., Reicher, J. J., Romera-Paredes, B., Sidebottom, R., Suleyman, M., Tse, D., Young, K. C., De Fauw, J., & Shetty, S. (2020). International evaluation of an AI system for breast cancer screening. *Nature, 577*, 89–94.
26. Pisano, E. D. (2020). AI shows promise for breast cancer screening. *Nature, 577*, 35–36.
27. Cohen, J. D., Li, L., Wang, Y., Thoburn, C., Afsari, B., Danilova, L., Douville, C., Javed, A. A., Wong, F., Mattox, A., Hruban, R. H., Wolfgang, C. L., Goggins, M. G., Dal Molin, M., Wang, T. L., Roden, R., Klein, A. P., Ptak, J., Dobbyn, L., Schaefer, J., Silliman, N., Popoli, M., Vogelstein, J. T., Browne, J. D., Schoen, R. E., Brand, R. E., Tie, J., Gibbs, P., Wong, H. L., Mansfield, A. S., Jen, J., Hanash, S. M., Falconi, M., Allen, P. J., Zhou, S., Bettegowda, C., Diaz, L. A., Jr., Tomasetti, C., Kinzler, K. W., Vogelstein, B., Lennon, A. M., & Papadopoulos, N. (2018). Detection and localization of surgically resectable cancers with a multi-analyte blood test. *Science, 59*, 926–930.
28. Zhavoronkov, A., Vanhaelen, Q., & Oprea, T. I. (2020). Will Artificial Intelligence for Drug Discovery Impact Clinical Pharmacology? *Clin Pharmacol Ther*. https://doi.org/10.1002/cpt.1795. [Epub ahead of print] Review.
29. Harrer, S., Shah, P., Antony, B., & Hu, J. (2019). Artificial Intelligence for Clinical Trial Design. *Trends in Pharmacological Sciences, 40*, 577–591.

30. Menden, M. P., Iorio, F., Garnett, M., McDermott, U., Benes, C. H., Ballester, P. J., & Saez-Rodriguez, J. (2013). Machine learning prediction of cancer cell sensitivity to drugs based on genomic and chemical properties. *PLoS One, 8*, e61318.
31. Rajkomar, A., Oren, E., Chen, K., Dai, A. M., Hajaj, N., Hardt, M., Liu, P. J., Liu, X., Marcus, J., Sun, M., Sundberg, P., Yee, H., Zhang, K., Zhang, Y., Flores, G., Duggan, G. E., Irvine, J., Le, Q., Litsch, K., Mossin, A., Tansuwan, J., Wang, W. J., Wilson, J., Ludwig, D., Volchenboum, S. L., Chou, K., Pearson, M., Madabushi, S., Shah, N. H., Butte, A. J., Howell, M. D., Cui, C., Corrado, G. S., & Dean, J. (2018). Scalable and accurate deep learning with electronic health records. *NPJ Digital Medicine, 1*, 18.
32. Freedman, DH. (2020). Hunting for new drugs with AI. *Scientific American,* S3–S7.
33. Lin, A., Giuliano, C. J., Palladino, A., John, K. M., Abramowicz, C., Yuan, M. L., Sausville, E. L., Lukow, D. A., Liu, L., Chait, A. R., Galluzzo, Z. C., Tucker, C., & Sheltzer, J. M. (2019). Off-target toxicity is a common mechanism of action of cancer drugs undergoing clinical trials. *Science Translational Medicine, 11*(509), pii: eaaw8412. https://doi.org/10.1126/scitranslmed.aaw8412
34. Palmer, A. C., & Sorger, P. K. (2017). Combination cancer therapy can confer benefit via patient-to-patient variability without drug additivity or synergy. *Cell, 171*, 1678–1691.
35. Rashid, M. B. M. A., Toh, T. B., Hooi, L., Silva, A., Zhang, Y., Tan, P. F., Teh, A. L., Karnani, N., Jha, S., Ho, C. M., Chng, W. J., Ho, D., & Chow, E. K. (2018). Optimizing drug combinations against multiple myeloma using a quadratic phenotypic optimization platform (QPOP). *Science Translational Medicine, 10*(453), eaan0941.
36. Wylliard, C. (2020). Can AI fix medical records? *Scientific American*, S13–S16.
37. John, E. M., Miron, A., Gong, G., et al. (2007). Prevalence of pathogenic BRCA1 mutation carriers in 5 US racial/ethnic groups. *Journal of the American Medical Association, 298*, 2869–2876.
38. Rebbeck, T. R. (2002). Prophylactic oophorectomy in BRCA1 and BRCA2 mutation carriers. *European Journal of Cancer, 38*(Suppl 6), S15–S17.
39. Saslow, D., Boetes, C., Burke, W., American Cancer Society Breast Cancer Advisory Group, et al. (2007). American Cancer Society guidelines for breast screening with MRI as an adjunct to mammography. *CA: A Cancer Journal for Clinicians, 57*, 75–89.
40. Domchek, S. M., Friebel, T. M., Singer, C. F., et al. (2010). Association of risk-reducing surgery in BRCA1 or BRCA2 mutation carriers with cancer risk and mortality. *Journal of the American Medical Association, 304*, 967–975.
41. Sénéchal, C., Reyal, F., Callet, N., et al. (2016). Hormonotherapy for breast cancer prevention: What about women with genetic predisposition to breast cancer? *Bulletin du Cancer, 103*, 273–281.
42. Menkiszak, J., Chudecka-Głaz, A., Gronwald, J., et al. (2016). Prophylactic salpingo-oophorectomy in BRCA1 mutation carriers and postoperative incidence of peritoneal and breast cancers. *Journal of Ovarian Research, 9*, 11. https://doi.org/10.1186/s13048-016-0220-4
43. Wang, Y., Cortez, D., Yazdi, P., et al. (2000). BASC, a super complex of BRCA1-associated proteins involved in the recognition and repair of aberrant DNA structures. *Genes & Development, 14*, 927–939.
44. Potapova, A., Hoffman, A. M., Godwin, A. K., Al-Saleem, T., & Cairns, P. (2008). Promoter hypermethylation of the *PALB2* susceptibility gene in inherited and sporadic breast and ovarian cancer. *Cancer Research, 68*, 998–1002.
45. Caminsky, N. G., Mucaki, E. J., Perri, A. M., et al. (2016). Prioritizing variants in complete hereditary breast and ovarian cancer genes in patients lacking known BRCA mutations. *Human Mutation, 37*, 640–652.
46. Russo, I. H., Koszalka, M., Gimotty, P. A., & Russo, J. (1990). Protective effect of chorionic gonadotropin on DMBA-induced mammary carcinogenesis. *British Journal of Cancer, 62*, 243–247.

47. Russo, I. H., Koszalka, M., & Russo, J. (1991). Comparative study of the influence of pregnancy and hormonal treatment on mammary carcinogenesis. *British Journal of Cancer, 64*, 481–484.
48. Russo, I. H., Koszalka, M., & Russo, J. (1990). Human chorionic gonadotropin and rat mammary cancer prevention. *Journal of the National Cancer Institute, 82*, 1286–1289.
49. Santucci-Pereira, J., George, C., Armiss, D., Russo, I. H., Vanegas, J. E., Sheriff, F., de Cicco, R. L., Su, Y., Russo, P. A., Bidinotto, L. T., & Russo, J. (2013). Mimicking pregnancy as a strategy for breast cancer prevention. *Breast Cancer Management, 2*, 283–294.
50. Peri, S., de Cicco, R. L., Santucci-Pereira, J., et al. (2012). Defining the genomic signature of the parous breast. *BMC Medical Genomics, 5*, 46.
51. Russo, J., Santucci-Pereira, J., de Cicco, R. L., et al. (2012). Pregnancy-induced chromatin remodeling in the breast of postmenopausal women. *International Journal of Cancer, 131*, 1059–1070.
52. Russo, J., Santucci-Pereira, J., & Russo, I. H. (2014). The genomic signature of breast cancer prevention. *Genes, 5*, 65–83.
53. Belitskaya-Levy, I., Zeleniuch-Jacquotte, A., Russo, J., et al. (2011). Characterization of a genomic signature of pregnancy identified in the breast. *Cancer Prevention Research, 4*, 1457–1464.
54. Liu, X., Holstege, H., van der Gulden, H., Treur-Mulder, M., Zevenhoven, J., Velds, A., Kerkhoven, R. M., van Vliet, M. H., Wessels, L. F., Peterse, J. L., Berns, A., & Jonkers, J. (2007). Somatic loss of BRCA1 and p53 in mice induces mammary tumors with features of human BRCA1-mutated basal-like breast cancer. *Proceedings of the National Academy of Sciences of the United States of America, 104*, 12111–12116.

Index

J. Russo, *The Future of Prevention and Treatment of Breast Cancer*,
https://doi.org/10.1007/978-3-030-72815-1

The manufacturer's authorised representative in the EU is Springer
Nature Customer Service Centre GmbH, Europaplatz 3, 69115 Heidelberg,
Germany. If you have any concerns regarding our products, please
contact ProductSafety@springernature.com

Printed and bound by CPI Group (UK) Ltd, Croydon, CR0 4YY
30/09/2026
02225694-0003